# RESISTING THE Green Dragon

## Dominion, Not Death

# RESISTING THE Green Dragon

## Dominion, Not Death

James Wanliss

Burke, VA: Cornwall Alliance for the Stewardship of Creation

**RESISTING THE Green Dragon: Dominion, Not Death**



Printed in the United States of America

Unless otherwise noted, all Scripture quotations are from The Holy Bible, New King James Version.

Cornwall Alliance for the Stewardships of Creation
9302-C Old Keene Mill Road
Burke, VA 22015
703-569-4653

Printed by
BookMasters, Inc.

ISBN-13: 978-1-886568-58-7
ISBN-10: 1-886568-58-8

Uroboros [Drawing by Theodoros Pelecanos, in alchemical tract titled Synosius (1478)].

The Ouroboros (Greek Ούροβόρος or ούρηβόρος, from ούροβόρος ὄφις "tail-devouring snake," is an ancient symbol depicting a serpent or dragon swallowing its own tail and forming a circle. (Wikipedia)

*For my beloved wife.*

# Table of Contents


Foreword .......... 13
Preface .......... 15

**CHAPTER ONE**
Green Dragon:
How Environmentalism Gets it Wrong and Endangers Society .......... 21
Introduction .......... 23
Death by Another Name .......... 25
Shifting Tides .......... 28
Cancer on the Planet .......... 28
Earth Day Follies .......... 33
Greens and Reds .......... 34
Green Religion .......... 37
Useless Breathers .......... 40
Conclusion .......... 42

**CHAPTER TWO**
The Church Complicit:
How Environmentalism Has Penetrated the Church
with Anti-Human and Anti-Christian Ideas .......... 45
The Blame Game .......... 47
Just Words .......... 50
John Dewey's Silly Sheep .......... 54
With the Force .......... 56
Greener Pastures .......... 57
Fat Free Religion .......... 58
Behind the Power (the Throne) .......... 62
The Church Complicit .......... 63
Going Green with God .......... 67
Greener than Thou .......... 69
Terrible Twins .......... 71
Ghostly Voices .......... 72
Ecofeminism .......... 74
The End of Mind .......... 76
Conclusion .......... 77

**CHAPTER THREE**
Created in the Image of God:
Why Humans Are More Special to God than Other Creatures .......... 81

The Beginning ........ 83
The God Who Is Present ........ 84
Deaf and Blind ........ 86
Dust of the Earth ........ 89
Reason ........ 93
Righteousness ........ 97
Holiness ........ 100
Broken Clay ........ 103
Grave to Grace ........ 105
Dirt and Dust ........ 107
Blood and Mud ........ 108
Grace and Glory ........ 110
Conclusion ........ 112

**CHAPTER FOUR**
Naked Ape:
How Nature is Not Divine, and Humans Are Spiritually
Superior to All Other Creatures ........ 115
Mountain Climbing ........ 117
Bad Karma ........ 118
Nature, Man, and Things ........ 120
Body Lice ........ 121
Heart of Darkness ........ 122
A Sacred Earth Charter ........ 125
Into Philistine Hands ........ 127
Cooked Goose ........ 132
Top of the Heap ........ 134
One with the Beef ........ 135
Far Out Cows ........ 138
The Cows of Eden ........ 139
Hope and Change ........ 142
Man and Beast ........ 143
Superior and Inferior ........ 144
Guilt Trip ........ 145
Death to Life ........ 147
Conclusion ........ 149

**CHAPTER FIVE**
Dominion:
How Humans' Filling and Ruling the Earth
Can Release it from Bondage ........ 151

The Source of Evil ........ 153
Man on Man ........ 156
One Blood ........ 157
Eco Imperialism ........ 160
Truth in Christ ........ 161
Self Image ........ 162
Holy Things ........ 163
Brother Sun, Sister Moon ........ 164
Hell in a Handbasket ........ 165
Cursed Cosmos ........ 167
God's Dominion ........ 168
My Clay Pots ........ 170
Dominion, Not Domination ........ 173
Fruitfulness ........ 176
Fig Trees Multiply ........ 178
Moaning and Groaning ........ 180
Conclusion ........ 183

**CHAPTER SIX**
Death Wish:
How the "Sustainability" Chimera is Harmful
to the Environment and Politically Dangerous ........ 187
Earth Ocean ........ 189
Butterfly Wings ........ 190
Sky Ocean ........ 192
Stewardship ........ 194
Kindness ........ 197
Mercy ........ 199
Balance ........ 200
Saving the World by Destroying It ........ 203
Groves and Graves ........ 205
Changing the Balance ........ 207
Collapse ........ 209
Nineveh ........ 211
Diversity ........ 213
Beautiful people ........ 218
The Holdren Bomb ........ 220
Multiply ........ 222
Invention of a Concept ........ 223
Conclusion ........ 227

**CHAPTER SEVEN**
Dominion, Not Death:
How Environmentalism Harms the Poor and Biblical
Christianity Offers the True Hope for the Future ........ 231
Snow Angels ........ 233
Know Pain, No Gain ........ 234
Blood and Gore ........ 236
Sweaty Palms ........ 238
Christianity as a Greek Cult ........ 239
Flesh ........ 243
Sprouts, Not Meat ........ 247
Der Kommissar ........ 252
Green with Envy ........ 256
Watermelons ........ 260
Heart Matters ........ 261
Sin ........ 263
Conclusion ........ 266

Endnotes ........ 272
Index ........ 302
Additional Resources ........ 312

## Foreword

If you want to understand the world view, theology, ethics, and politics of the Green movement, and you want those views subjected to a penetrating critique by a Biblically and theologically astute Christian with impeccable scientific credentials, this book is for you. *Resisting the Green Dragon* will not only help you understand atheistic, pantheistic, and animist environmentalism, but also help you define the nagging sense you've had that there's not only a lot right but also *something* wrong with the attractive, Christianized version of environmentalism that calls itself *creation care.*

When many Christians think it pious to uncritically embrace doom-and-gloom ecology and demands for slowed economic development and growth, reduced consumption, and government-imposed austerity, James Wanliss shows how they compromise the Biblical faith and promote unbiblical asceticism supported by poor science and flawed economics leading to policies that oppress the poor while enriching the powerful.

But much more important, he outlines a clear, Biblical understanding of Earth stewardship—the fulfillment of God's first command to humanity, the Dominion Mandate to "be fruitful and multiply; fill the earth and subdue it; have dominion over the fish of the sea, over the birds of the air, and over every living thing that moves on the earth" (Genesis 1:28). At the same time he clearly and faithfully proclaims the gospel of reconciliation with God for sinners justified by grace alone through faith alone in Christ alone, thus obeying Christ's last command on Earth, the Great Commission to "make disciples of all nations . . . teaching them to observe all things that I have commanded you" (Matthew 28:19–20).

Sincere Christians searching for a clear understanding of how to be good stewards of God's Earth while loving their neighbors and faithfully representing Jesus Christ and His glorious gospel will find *Resisting the Green Dragon* irresistible!

E. Calvin Beisner, Ph.D.
National Spokesman, The Cornwall Alliance for the Stewardship of Creation
Author, *Where Garden Meets Wilderness: Evangelical Entry into the Environmental Debate* and *Prospects for Growth: A Biblical View of Population, Resources, and the Future*

## Preface

In 2007, at my former university, I taught a course entitled "*The Politics and Science of Fear*." My motivation was a desire to explore areas of science that elicit irrational fear and loathing. Our first topic was energy. One of the safest and cleanest energy sources is nuclear power, but it often elicits hysterical behavior from the Green movement and much of the media. Next we considered organic foods, pesticide use, water quality, and finally global warming—all environmentalist shibboleths. We even watched the documentary *End of Suburbia*, an ode to the passing of Western civilization.

My own interest in the Green movement was stirred in the fall of 2006 at the American Geophysical Union professional meeting. Here former U.S. Vice President Albert Gore presented his theory of global warming. Given the recent success of his scientifically fallacious, but Hollywood acclaimed, global warming documentary I considered it odd that he received an invite to talk to one of the world's premier scientific meetings. Surely his many exaggerations and errors would disqualify him. Yet the crowd was delirious with pleasure, which shows at the least that many thousands of nerdy scientists are quite as capable of trendy groupthink as lesser mortals. Afterward it was more than a little troubling to watch the enormous crowd slowly uncoil itself, like a human dragon.

Gore's presentation at the meeting, plus the prominent presence of the Union of Concerned Scientists, a highly partisan environmental advocacy organization, suggested to me that my area of science was being exploited for political purposes. When my university showed Gore's documentary around twenty times during March 2007, I was close to fed up. Environmentalist propaganda pressed on every side, and I became more determined than ever to offer balanced alternatives to my students.

On reaching the section on global warming in the course, we watched two films. One was *The Great Global Warming Swindle*, and the other Gore's *An Inconvenient Truth*, just in case any of my students had missed the previous campus showings. The impetus to write this book was sealed when one horrified physics faculty member suggested that to show *Swindle* was irresponsible and akin to teaching Nazi propaganda!

It is, sadly, increasingly accurate to characterize Western universities by intolerant political correctness, that woeful ideology intent on taming unbridled conversation. This seems to be well-known outside academia. So when the local newspaper heard the story of a college professor questioning campus orthodoxy, it thought it worth a story. The article was sufficiently atypical that the most used news website on the internet, the *Drudge Report*,[i] picked it up. These events contributed to the birth of this book.

What *Resisting the Green Dragon* exposes, and challenges, is the religious machinery that forms the veiled foundations of environmentalism. The book details the values of the environmental movement and contrasts these with the Christian virtues that built Western Civilization. It seeks to display the incompatibility of modern environmentalism with that Civilization and suggests an alternative path to dominion, not death. The Christian community must be part of the solution but instead often abounds with unbiblical views of man and nature. This book explains how the Green movement seeks to co-opt the church. It shows how and why the church should resist and offers suggestions to the proper relationship between humanity and the rest of creation.

Virtually every day there are demands for people to Save the Planet; guilt manipulators are in their element. It would be good to shake free of this Green haze, to get a breath of fresh air. What we all need is an honest and prayerful evaluation, or re-evaluation, of the character of the modern environmentalist movement. This is a difficult proposal, since we are constantly bombarded with emotional appeals to be "Green," whatever that may mean. And if we are not "Green" they say the world will end. What must sincere Christians and others of sensitive and caring nature do?

Yet in spite of the doom saying, human welfare in a host of categories—including life expectancy, health, education, wealth, and so on—is actually improving. Environmental quality, at least in societies most strongly influenced by Christian values, continues to improve. However, this book is not about direct measurements of human or environmental welfare. I do not provide scientific or economic

[i]http://www.drudgereportarchives.com:data:2007:05:13:20070513_005846.htm; last viewed on 9/21/2009.

answers to the Green trope that the world is ending because of human greed and population growth. Multiple excellent resources already do that job, many of which appear in the endnotes.

So what should Christians make of the Green movement? Certainly, the dues-paying rank-and-file members of environmentalist groups are probably well-meaning people of goodwill. Many profess faith in Jesus Christ. They care deeply about the world in which they live. Unfortunately, in our age of fast-moving pictures, the public image of environmentalism easily dupes softhearted folk. Fuzzy bunnies, baby polar bears, attractive celebrities all are levers that enjoy powers of emotional persuasion. But the disguise must be peeled away. Christians are to be bloodwashed, not brainwashed by fads of secular society. Seven chapters are devoted to exposing a movement I have called the *Green Dragon*, after what I argue is its diabolical original. Chapter 1 reveals the agenda of the environmentalist movement. Perhaps surprisingly, that agenda has precious little to do with the natural environment. Where possible I have quoted the words of leaders and trend makers in environmentalism—they speak for themselves and show how their claim to care about the environment seems more of a mask for something unpleasant. From the writings and public declarations of prominent environmentalists it is clear the ethos of the Green movement fosters deep animus and hostility towards humanity. What environmentalist leaders have to say about them will shock caring folk who value both fidelity and the natural environment.

Chapter 2 explores attempts being made to infuse the Christian Church with an environmentalist reinterpretation of the Gospel of Jesus Christ. Who wants a toxic waste dump in their backyard? Not Christians. Not environmentalists. But their reasons for this conclusion are quite different. Environmentalism has penetrated the church with anti-human and anti-Christian ideas. Rather than rebuffing environmentalist tactics, the church has shown herself remarkably susceptible to green guilt manipulation. This Chapter shows how environmentalists exploit religious language with the hope of manipulating the religious response into a changed consciousness and biospheric political activism. They wish the dynamism of religious devotion, for which many are willing to die, would give impetus to their visions of a new world order.

Chapter 3 deals with the Christian understanding of the image of God. It contrasts this view with the standard vision adopted by the Green movement, which holds that our species is nothing special in the natural scheme of things. Greens see similarities between humans and beasts, but fail to see that spirit elevates man above the beasts. The Green movement's radical departure from truth is logical. It arises from the Greens' faith in the idea that humankind has animal ancestry. But in fact, humans are quite different, and more special to God than all other creatures. Whether Greens like it or not, human life is sacred, and infinitely precious because made in the image of God. Animals are not.

Chapter 4 pushes the Green ethos back still further. If men will not worship God, they will worship what God has made, for worship they must. And if forced to recognize that man is special, Greens do more than wonder that nature is too. Chapter 4 crushes this idea with the teaching of the Word of God. Nature is special, but only as the stage on which Christ's redemption of humanity unfolds. Human beings are not only *very* special, but spiritually superior to any other part of the created order. Unfortunately, many in the professing church argue that Nature is divine. Nature is not only *not* divine, it is also not eternal. It is subject to mankind.

Chapter 5 continues to consider our position in the natural world, and our spiritual relationship to that world. It explains the Biblical doctrine of dominion. Scripture gives a faithful, realistic, and tender presentation of the sad effects of human frailty and sin on the entire universe, and our responsibility for Biblical stewardship and dominion. That people dominate the landscape oppresses environmentalists. The idea that humans *should* take dominion is, to them, always an abomination. Environmentalists dismiss, and Christian Greens underestimate, the moral and ethical consequences to the Earth of the sin and curse described in Genesis. Rather than humankind being Earth destroyers, it turns out that only our filling and ruling the Earth to God's glory can release it from the Genesis curse. All of history spirals towards this one point, that God has kept the human race to prove His special care for His church. Therein lies hope for the entire cosmos.

Chapter 6 addresses the idea of "sustainability," a thinly veiled political project, based on the mythical balance of nature and disturbance of that fragile "balance" through the energetic, fruitful activity of humans. The Green accusation is that rather than waiting patiently for Gaia's gifts humans force the Earth. Instead of waiting in humble dependence, humans irrigate. Instead of waiting for food to drop off trees into their mouths, they cultivate. Sustainability argues that this is unrighteous. But sustainability ignores God's hand in history. This chapter exposes "sustainability" as a code word for Green goals of wrenching, harmful changes in social institutions, economic structure, and, finally, human consciousness. The sustainable world envisaged by environmentalists features command-and-control economics that, contrary to the Biblical emphasis, places human life directly in the crosshairs of violent men.

Chapter 7 considers a few prominent demands of the Green movement and its false accusations against Biblical Christianity. Greens want to control how you live, eat, drive, and even the light you use to read by. Environmentalism is a dream come true for busybodies who want to micromanage the details of the lives of others, all to protect the Earth, at all costs, from human interference. It boils down to the utopian idea that there can be Heaven on Earth. It confuses God's kingdom with pagan ideas to try to "save" the world. Its slogan is planetary salvation, but its way of salvation is dangerously out of touch with the most basic of human needs and emotions. It is ultimately dreadfully harmful to the environment and humans, particularly the poor. Chapter 7 closes with an appeal to the reader—stop supporting a Green Dragon that wants to strike you down. Proclaim the only true hope for humankind and the entire universe—the unadulterated gospel of Jesus Christ. Resist the Green Dragon! Rise up and strike off its head!

*James Wanliss, Seoul, South Korea, July 2010*

***Ex Afrika semper aliquod altior***

# CHAPTER ONE

# Green Dragon: How Environmentalism Gets it Wrong and Endangers Society

*Everyone's freaking out about economic collapse; meanwhile, the oceans are silently collapsing and no one's really doing anything about it.*

Daryl Hannah[1]

## Introduction

While many nations now enjoy prosperity unprecedented in history, for many people in those nations the early twenty-first century brings with it freezing despair and lack of confidence that this prosperity can continue much longer. Increasingly these days there have been louder cries that societal collapse may come due to destruction of the environment on which we depend. From 2001 to 2008 more and more Americans, to a high of sixty-eight percent of those in a 2008 Gallup poll, were pessimistic about the future and said the quality of the environment was getting worse.[2] A *Pew* survey of 47 nations found a general increase in the percentage of people citing pollution and environmental problems as a top global threat.[3] Given the overwhelming media prevalence of what Bjørn Lomborg calls "The Litany,"[4] maybe it should not be surprising that people feel this way, whether in America, Europe, or any of the developed nations.

"We are all familiar," says Lomborg, "with the litany of our ever-deteriorating environment. It is the doomsday message endlessly repeated by the media, as when *Time* magazine tells us that "everyone knows the planet is in bad shape," and when the *New Scientist* calls its environmental overview 'self-destruct.' We have heard the litany so often that yet another repetition is, well, almost reassuring."[5]

Every traveler likes to throw a stone where a pile already sits. It gives one a satisfying sense of being part of an historical moment, of fitting in, of making a contribution. Could it be that the renewed urgency with which ecological issues have intruded into human consciousness has more to do with the need to fit in and have a bit of peace than any actual ecological crisis? But it is hard to think beyond the clanging and relentless media deluge. As perception of danger increases people will worry more, leading to more enthusiasm for reporting the concern, and so to more fear.

Even indifferent folks cannot fail to notice the ubiquitous nature of environmentalist propaganda. *Ad nauseam* updates on billboards, television, and radio trumpet peace prizes, movie awards, and other accolades awarded former U.S. Vice President Al Gore, who has become something of an environmentalist rock star. He must find it gratifying to have received entertainment industry recognition in an Oscar. Genuine rock stars can only fantasize about such notoriety.

Of course, the Litany is not really new. Al Gore has preached variations on this theme for most of his political career. For example, in 1992 he wrote: "The disharmony in our relationship to the earth, which stems in part from our addiction to a pattern of consuming ever-larger quantities of the resources of the earth, is now manifest in successive crises, each marking a more destructive clash between our civilization and the natural world.... The loss of one and a half acres of rain forest every second, the sudden, thousandfold acceleration of the natural extinction rate ... the ozone hole ... the possible destruction of the climate balance that makes our earth livable—all these suggest the increasingly violent collision between human civilization and the natural world."[6]

Although not uncommon in fringe literature, this rather breathless kind of synopsis was still far out of the mainstream in the early 1990s and threatened Gore's political ambitions. Now it is the mainstream, and beyond politically correct; environmentalism is suprapolitical. American politicians from both Republican and Democratic Parties support radical environmental policies.

If it were merely a peccadillo of a marginalized minority, then perhaps one should spend little effort refuting the premises expressed in the Litany. But the truth is that the Litany now permeates society at every level. Also, many people of good character subscribe to the Litany. That there are true believers should not surprise, given the obvious and extensive efforts made to indoctrinate the public and to discourage critical analysis of the Litany. To decline meek obedience, as Lomborg has done, is unfortunately to risk professional ostracization or worse.[7] Under the oppressive burden of such conditions it is difficult enough to question, let alone resist actively, an idea taught to the exclusion of others from kindergarten through university, printed in newspapers and magazines, and seen and heard multiple times each day on television and radio.

But it is important to question and challenge the Litany. It deceives too many well-intentioned people. Not only is the Litany mistaken,[8] but, as we will see, its deep religious foundations are in fundamental conflict with truth. Christianity is a religion professing the greatest affinity for truth, Jesus even calling Himself "the truth." Christians must care, therefore, that they do not misguidedly accommodate the world spirit of this age under the guise of caring for the environment.

Now it would be easy to set up an easel, spread a canvas on it, and paint environmentalists in the darkest colors, imputing to them only evils and abuses. This is how many arguments proceed. Or one *could* list all the scientific falsehoods we can find in the Green framework. This is an important and useful tactic, but the list will have no end and will not end the debate between those who care. The argument here is different. The environmentalists will speak for themselves.

In the words of theologian Robert Dabney, "He who prunes the offshoots of error has an endless task; a task which usually results only in surrounding himself with a thicket of thorny rubbish. It is better to strike at the main root of the evil stock from which this endless outgrowth sprouts."[9] It is our goal, therefore, to strike at the root from which the Litany has sprung. But first one must step back and survey the growth, to see if its fruits are as evil as suggested, then seek the root and destroy it.

## Death by Another Name

The Litany has been consistently preached by various visionaries and scientists for a long time, providing newsworthy fodder. In 1970, at the first *Earth Day*, Stanford University biology Professor Paul Ehrlich claimed, "In ten years all important animal life in the sea will be extinct. Large areas of coastline will have to be evacuated because of the stench of dead fish."[10] Throughout the 1970s big media conglomerates trumpeted this line from the Litany, and other scares such as the new age of global cooling. They claimed that global cooling would cause "droughts, floods, extended dry spells, long freezes, delayed monsoons, and even local temperature increases…."[11] Yes, Virginia, all this from global cooling. The *New York Times* regularly chimed in with headlines such as, "Climate Changes Endanger World's Food Output."[12] Even the U.S. Central Intelligence Agency got a piece of the action, with *US News and World Report* providing details from the CIA investigation: "the earth is heading into some sort of mini-ice age."[13]

Big media has not changed its way of working—if it bleeds it leads—and unfortunately obliging scientists have not been hard to find. While in the 1970s it was global cooling and another ice age that supposedly threatened humanity, today global warming hysteria feeds the headlines. Following an episode of global warming fear and the concomitant "Next Great Deluge Forecast by Science,"[14] there was some hope the end of the world might be delayed just a little longer. A 1952 headline read, "Rising World Heat Seen Leveling Off."[15] But in 1923 headlines warned of the "Menace of a New Ice Age."[16] So within a short time warnings from scientists and the media repeatedly see-sawed between global warming and global cooling cataclysms.

Stories about ice ages had been all the rage even in 1895 when geologists had first discovered the remains of forests of tropical plants in frozen Greenland.[17] Given this apparent predilection to swinging between stories of global warming and cooling, it was only natural that in 1890 the reported "consensus" of scientists was that New Yorkers would face a balmy few thousand years.[18] And now, in the twenty-first century, we hear of an imminent threat of the end of the world, once again because of excessive warming. An apocalypse served up alternately frozen or fried, with little Green garnishes on the theme, is not so appetizing. We are as a race, journalists included it would seem, prone to catching the nightmare whimsies of the weather scarecrows.

All of this is not to say ecological disasters do not occur. We wish they did not, but, of course, they do. Neither does this mean that people don't behave badly or always do what is right. The history of the world is one train wreck after another of bad behavior and disasters, those that have been man-made or man-induced, and those that have emerged out of the blue. One ancient example of climatic and ecological cataclysm is the Great Deluge (Genesis 7:24) that overflowed the entire Earth and wiped out virtually every living creature. Thankfully we have a promise that God will never again express wrath like this against sinful mankind. Al Gore acknowledges, "It was the promise never again to destroy the earth by floods, which, according to Genesis, is the symbolic message of every rainbow."[19]

Another ecological disaster, though not so great in extent as the flood, was the monstrous famine that occurred roughly 1,750 years before Christ (Genesis 47:13), which persuaded the family of Israel to journey to Egypt, where there was food. Long after those days, in 42 AD, during the second year of the reign of Claudius Caesar, a severe famine raged at Rome, followed by regional famine that began in the fourth year of his reign.[20] Little had changed over a thousand years later. Julia Marvin writes: "From 1315 to 1318 England—and much of Europe—suffered three rainy years and six ruined harvests in a row. By 1316 famine had begun. Seed-corn rotted in the damp, mass malnutrition led to death from disease, and the desperate resorted to crime, all in a time of war."[21]

During the twentieth century, arguably the bloodiest century of all, Stalin and Mao managed to devastate the land and starve around 40 million humans. Another 70 million lives became lost through the direct actions of the various Soviet "cooperatives,"[22] all in peacetime. World War I and the subsequent monomania of Hitler and his financiers pushed nearly another 100 million out of this world. History is the sorry harvest of multitudes trying to dodge tyrannies, wars, diseases, and famines. Thomas Hobbes got to the root of it when he succinctly summarized reasonable human expectations: "… the life of man, solitary, poore, nasty, brutish, and short."[23] History teaches that, for most folks, "The Good Old Days–They Were Terrible!"[24] Plague, war, and other intermittent crises arising from combinations of political, ecological, and climate changes stain and scar the annals of the world like somber gravestones punching through the bleeding earth.[25]

Only since the Renaissance and Reformation of the fifteenth through seventeenth centuries, and the Industrial Revolution starting in the eighteenth, has mankind begun to exercise anything like effective dominion over the terrors of restless Earth. These events brought breakthroughs in political and economic order, science, and technology by which nature was disenchanted, paganism receded, and innovations of every kind multiplied exponentially. For the first time in history economic growth occurred rapidly enough to see. Those countries strongly affected by the Protestant Reformation especially enjoyed improved quality of life. Life expectancy rose from about 30 years in the 1400s to about 80 years by the year 2000.[26]

Near the time of the new millennium, and for just a brief, yet hilarious, stretch we came to enjoy a different terror. Some friends are still eating beans they had stored, "just in case." Death by PC meltdown was a pleasant and transient respite from the usual suspects. But even during this time reminders came from the United States Deputy Secretary of Defense John Hamre that environmental terror was, after all, only temporarily on the backburner: "The Y2K problem is the electronic equivalent of the El Niño and there will be nasty surprises around the globe."[27] There is only one Y2K, but death by weather is certain.

## Shifting Tides

Mass media influence alone cannot explain the cosmic shift towards a social climate that begins to favor the environment—polar bears, trees, and bugs—over human beings. The shift in atmosphere is not a recent phenomenon. Western elites have been proclaiming and preaching this for a long time, with varying degrees of zeal and varying degrees of success in capturing the public ear. Born in 1712 and raised in Calvinist Geneva, Jean-Jacques Rousseau was one such successful apostle who preached that man is innately good but has embraced evil by avoiding "nature" and becoming "civilized." "God makes all things good," wrote Rousseau, "man meddles with them and they become evil."[28] While much of Europe has imbibed from this troubled spring, it is arguably only in recent times, with the increasing pace and ubiquity of government, and with the nearly monolithic message of the mass media with its dominance as a source of information, that ideas like this have gained significant cultural traction. An unremitting societal shift away from Christian thinking has only aided this transformation.

What we usually denominate as Western civilization is largely Judeo-Christian civilization. The massive advancements of recent centuries sprang chiefly out of Christian belief and principally from the practice of the Reformation. For all its faults, the great benefits and attainments of Western civilization are the result of generation after generation seeking to act out Christian principles, to answer the most basic questions of all—who are we, and how should we then live?[29]

Yet the fruits of this worldview, for example decreasing human mortality rates, do not always receive applause, and less in these days than before. From Agricola[30] to Rousseau to Malthus, and beyond, Western elites have consistently lamented the contribution of Earth's abundance to the increasing human life expectancy.

## Cancer on the Planet

To the radical environmentalist Western civilization is "...a nightmarish chronicle of exploitation, thievery and environmental degradation. For Western elites, the earth was a Garden of Eden until corrupted by Western imperialism and capitalism, not to mention Western religion…."[31] According to Nietzsche's fire-dog, "The earth hath a skin; and this skin hath diseases. One of these diseases, for example, is called 'man'."[32] Humans (Americans in particular) are thus a "cancer" on the planet that must be cut out, hopelessly breeding out of control, cutting, killing, poisoning and polluting, and to top it all off, slow cooking the skin of the Earth.

A friend of the philosopher David Hume, and acquaintance of Rousseau, explained why people are so destructive. Rev. Robert Malthus first taught that this naughty behavior was the unfortunate, but inevitable, result of allowing what he considered the lower class of human beings to breed unrestrained. Malthus, whose calling was as an Anglican preacher, seems to have learned more on the knobbly knees of Hume and Rousseau than from the Bible. He wrote: "The power of population is so superior to the power of the earth to produce subsistence for man, that premature death must in some shape or other visit the human race. The vices of mankind are active and able ministers of depopulation. They are the precursors in the great army of destruction, and often finish the dreadful work themselves. But should they fail in this war of extermination, sickly seasons, epidemics, pestilence, and plague advance in terrific array, and sweep off their thousands and tens of thousands. Should success be still incomplete, gigantic inevitable famine stalks in the rear, and with one mighty blow levels the population with the food of the world."[33]

In his view humanity's spectacular success spells doom. Any improvement in technology to ease the human condition would lead to increased breeding by lower classes, from here to overpopulation, to suffering, and on and on in a pathetic and pitiless cycle.

Malthus favored moral restraint as a check on population growth but proposed this only for the lower classes, then known as the breeding classes. While the ideas he promoted appeared theoretically compelling, they have never been observed in history. In fact, for sheer body count the two cruelest famines in history were fomented by vile theories of atheistic governments, not uncontrolled breeding. In the Soviet Union (1932–1933) there were 7 to 10 million Ukrainian[34] victims, and in Communist China (1958–1961) there were above 30 million victims.[35] To say, as some do, that these famines arrived from too many useless eaters would be like saying that the National Socialist (Nazi) holocaust rained on man because of too many Jewish meteorologists.

Poverty, not people, is a pollution. Historical data shows Malthusian ideas are a crock, perhaps because they fail to grant that whereas human population may at best increase geometrically, human innovation, if nourished by freedom, increases in a much faster power law fashion.[36,37] As noted by Soichino Honda, the founder of the Japanese motor company bearing his name, "Where 100 people think, there are 100 powers; if 1,000 people think, there are 1,000 powers."[38] This is an important insight, but one unknown to Malthusians since they fail to discriminate the great traits that separate the human from the animal. To put it another way, man and coyote both eat chickens. But the more coyotes the fewer chickens, while the more men the more chickens. This is certainly the case in most countries strongly influenced by Christian principles. There a single genius can bless the destiny of the whole country, or even the whole world, but a pretty monkey has never even asked the question.

After strong early resistance, in the 1800s Malthusian ideals quickly gained currency, first in England and then Europe. Charles Darwin cited[39] Malthus as a key forerunner of his own most famous and influential work, entitled *The Origin of Species by Natural Selection; or, the Preservation of Favoured Races in the Struggle for Life*.[40] Then, for a while, it was all the rage to speak of the inevitable destruction of various natural resources along the lines of the Malthusian thesis. Thinkers believed it slightly less than credible to "imagine a time when England will be a mere name in history"[41] because of the population growth of the breeding classes, and the declining natural resources, coal in particular.

In 1898 the luscious problem of expansion in the horse population, especially their horsey digestions, exercised urban planners. According to Eric Morris, "One New York prognosticator of the 1890s concluded that by 1930 the horse droppings would rise to Manhattan's third-story windows. A public health and sanitation crisis of almost unimaginable dimensions loomed."[42] Although these and other prominent Malthusian predictions have been discredited by spectacular failure, the intelligentsia still cling to the thesis.

Twenty-first century environmentalists hold feverishly to the excruciatingly bleak Malthusian vision. Malthus predicted environmental degradation, populations breeding like bacteria in a Petri dish, and a plethora of other self-inflicted disasters. He was wrong. All his disciples have been wrong. But his "solutions" never go out of fashion; economic collapse is preferred, as in the October 2008 issue of *New Scientist* magazine, whose cover read, "The Folly of Growth: How to stop the economy killing the planet." The answer is, of course, statist controls plus no more growth, for "if we are serious about saving Earth, we must reshape the economy."[43] Never mind the horrendous environmental record of command-and-control economies; neo-Malthusians require masochism from the masses. When coal in England was presumed to be near exhaustion, Malthusians called for a lower standard of living and industrial decay, because "We have to make the momentous choice between brief greatness and longer continued mediocrity."[44]

A century or so later, in a global era with more grandiose pretentions, it is more than England that is expected to fade away, but the entire human species, except perhaps for those that move to Mars. According to a 2006 Distinguished Texas Scientist: "I think we should take the Library of Congress up with us on DVDs and so when we wink out in this little sphere, there will be a little bit of a record of what happened on Earth somewhere else. And I think in that new plan, the books kids read in kindergarten will say 'The Rape of the Earth, Let's treat our planet a little better.'"[45]

Older pessimists, on balance, did not doubt the special place of mankind in the world and usually put humans first, or at the least some humans first. Nowadays, and except for the regrettably rare exceptions, coming environmental ecopalypse[46] is claimed to be caused by humans in general and Americans in particular. The silver lining to radical environmentalists, and others who fundamentally loathe their own race and civilization, is that humankind will receive its comeuppance for its rapine of Gaia, our Earth Mother.

John Davis from the group Earth First! once expressed rather quaintly the logical conclusion of the radical environmentalist ethos: "Human beings, as a species, have no more value than slugs."[47] What fascinates is that he was likely saying this for effect. Probably he believes that slugs have more intrinsic value than humans—this is after all a reasonable and mature final resting place for views springing from and through neo-Malthusian and neo-Darwinian perspectives. According to David Graber, a biologist with the United States National Park Service, "Human happiness and certainly human fecundity are not as important as a wild and healthy planet. I know social scientists remind me that people are part of nature, but it isn't true…. We have become a plague upon ourselves and upon Earth. Until such time as *homo sapiens* should desire to rejoin nature some of us can only hope for the right virus to come along."[48]

When they are not growing misty eyed about the prospect of pandemics wiping out mankind they turn to pejoratives, such as comparing humans to a virus. For example, Sir James Lovelock, a genius whose heart is sometimes in the right place, combined both the cancer and the virus in one piquant verbal poison pill: "Humans on the Earth behave in some ways like a pathogenic micro-organism, or like the cells of a tumor."[49] Similar unkind pronouncements flow endlessly, not only from fringe freaks, but also from influential people such as educators, politicians, royalty, and business leaders, to list a few. Radical environmentalism is no longer the fringe belief of a few powerless lunatics; it is mainstream. Many words of environmentalists suggest that they are not so much for the natural environment as they are against human life.

James Cameron's 2009 sci-fi extravaganza, *Avatar*, suggests that humans are a kind of universal filth, a plague afflicting the entire cosmos. The *Avatar* message echoes a scene, a decade earlier, from the wildly popular movie *The Matrix*. There the villainous cyberentity Agent Smith interrogates one of the heroes. He strokes beaded sweat off the bald head of his enemy, like rain that drips off waxed fruit, and says, "Every mammal on this planet instinctively develops a natural equilibrium with the surrounding environment. But you humans do not. You move to an area and you multiply and multiply until every natural resource is consumed and the only way you can survive is to spread to another area. There is another organism on this planet that follows the same pattern. Do you know what it is? A virus. Human beings are a disease, a cancer of this planet. You are a plague. And we are … the cure."[50]

This is something of a pointer to what the Green Dragon that is radical environmentalism wishes for the human race. To be sure, Prince Philip, the consort to the Queen of England and President Emeritus of the World Wide Fund for Nature (WWF) makes it a little clearer: "In the event that I am reincarnated, I would like to return as a deadly virus, in order to contribute something to solve overpopulation."[51] Humor evaporates. He would weep and go yogic for the death of a tree, yet he pines over personally seeing to human genocide. Paul Ehrlich was way ahead of the curve, writing over four decades ago, "A cancer is an uncontrolled multiplication of cells; the population explosion is an uncontrolled multiplication of people. We must shift our efforts from treatment of the symptoms to the cutting out of the cancer."[52]

Ehrlich recognizes that his murky dream world can only be realized through the business end of a gun, but he does not shrink from this conclusion. "Coercion?" says he, speaking with relish of his dreams for forced sterilization programs in India. "Perhaps, but coercion in a good cause."[53] One wonders what holds these genteel folk back. Perhaps just power—but they are working on that. Any rational person raised in the residue of Christian civilization should recoil from these radical ecotopian visions.

## Earth Day Follies

Not every environmentalist is so honest or outspoken, and most have not been so consistent to think their dogma through to this conclusion. Many are tilting at windmills and genuinely grieve their vision of just a few breeding pairs of humans surviving the ecopalypse. But the consistent environmentalists are increasingly coming out of the dark woodwork, a discharge infecting the bloodstream, like some grimy parasite inflamed on the hot blood of human progress. Only a razor thin porous membrane divides these hard core environmentalist ideologues from the usual soft and cuddly media portraits of the movement.

Earth Day is the premier environmentalist farrago. The US government promotes this brand of eco-spirituality with surprising vigor through tax funded schools and cooperative efforts across multiple government agencies.[54] Ostensibly intending to inspire awareness of and appreciation for the Earth's environment, Greens observe Earth Day yearly on April 22, an inauspicious day shared with the birth of

Vladimir Lenin, a Red Soviet genocidal enthusiast. In 1970, at the first Earth Day, there were supposedly only five to ten years left to begin taking action to lessen population growth and reverse the damage humans had done to Mother Earth.

Four decades on, Greens still urge society to make sacrifices and take immediate and drastic action, for the end is still nigh. *Time Magazine* proclaims, "At the start of the $21^{st}$ century there were unmistakable signs that exploitation of the planet was reaching its limit—that nature was beginning to take its revenge."[55] It seems that Earth's ecosystem is *always* supposedly at a tipping point. Indeed, influential politicians and wealthy businessmen point out that "Each passing day brings yet more evidence that we are now facing a planetary emergency—a climate crisis that demands immediate action to sharply reduce carbon dioxide emissions worldwide in order to turn down the earth's thermostat and avert catastrophe."[56]

Smug millennial politicians from both Republican and Democratic Parties in the United States support radical environmental policies. In a 2008 interview with the *San Francisco Chronicle*, President-elect Barack Obama said that his contribution to avert a climate crisis will be to "bankrupt" coal fired power plants. Given that coal is the fuel that produces over fifty percent of American energy, the little people will have to sacrifice and feel pain—a fact Obama was honest enough to admit. "Under my plan of a cap and trade system," said he, "electricity rates would necessarily skyrocket."[57] Meanwhile saner nations continue to build power plants. In communist China energy prices keep on going down as they build six coal-fired power plants each month.[58]

## Greens and Reds

Faith in environmentalism has turned to the point where it needs one to believe the most egregious insult to the Earth is the constant human production of carbon dioxide, a fundamental plant food. Each adult produces about 2.2 pounds per day simply by breathing, not including energy use for other acts. For this environmentalists encourage humans to harbor guilt and anxiety.

Those that keep fit and exercise naturally breathe out more than if they remained still. Finally fat people have something to celebrate.

They are carbon sinks, and they tend to die younger, not like those skinny useless breathers that pollute the world with their exercising, their selfish heavy breathing. But just today I learned it may not be so simple an equation, after all. Scientists and physicians have made this startling discovery: "It is true fat people eat more food than average." Who would have thought it? After extensive research and statistical analysis Phil Edwards of the London School of Hygiene and Tropical Medicine came to the following conclusion, "Moving about in a heavy body is like driving in a gas-guzzler."[59] According to his calculus, if we stop excessive food production and consumption we can immediately cut down so-called greenhouse gases by twenty percent (but savings are potentially much more if we factor in the resulting famines).

We are a plague on our houses and on the Earth, and our very life dooms us to dying—so the environmentalist mantra. To an already fearful people whose fevered dreams are built on sands of secularism, this new paradigm is the sum of all fears. It embodies in it all the hurricanes, tornadoes, floods, droughts, plagues, pestilence, pollutions, and almost any other ecological disaster that one can imagine. You name it, and global warming will claim it. In *The Hitchiker's Guide to the Galaxy*, a spiritually ambiguous people seek the answer to the Ultimate Question of Life, the Universe, and Everything. After eons of calculation the answer proposed is the number 42, which in postmodern parlance obviously means the same as "global warming." At least so says the United Nations Intergovernmental Panel on Climate Change (IPCC).

What a splendid paradigm—it is a catchall, seemingly able to explain anything from the blasé to the banal. Who would have thought Bulgarian brothels[60] would suffer barren business because of balmy weather? It seems Greens leave no bizarre brick unturned; this greatest of environmentalist peccadilloes will not only lengthen the day[61] by slowing Earth's rotation, but has already apparently caused the death of the Loch Ness Monster,[62] increased Italian suicide rates,[63] and birthed two-headed turtles.[64]

Bear in mind that as of 2009 the net warming anomaly the alarmists claim causes these upheavals amounts to a minuscule 0.6°C in 100 years. Forget about global warming, the *real* danger in a problem that affects every area of life is that the solution will also presumably need to invade every area of life. The creature of Loch Ness may be a legend, but a real monster is breeding right under our collective nose. How, in a professed age of science, are we being terrified by pseudo-science, even though packaged by expert pied pipers.

"How extraordinary!" said the political scientist Aaron Wildavsky, "The richest, longest-lived, best-protected, most resourceful civilization, with the highest degree of insight into its own technology, is on the way to becoming the most frightened."[65] It may not be too much of an *en passant* to rationalize the increasing suicide rates among Western youth in terms of the dissonance the supposed environmental crisis foments. If it is true that human beings are not merely insignificant, but the root of the problem, then the final solution to save the Earth is human extinction. The price to join that club is steep.[66]

Environmentalism breeds among its followers rage, guilt, angst, and fear. But in an age where the Christian basis for Western civilization is by turns ignored, neglected, twisted, resisted, and reviled, the Litany of the Green Dragon also provides some certainty for people without God, who drift steadily from their rational moorings, and for whom there is an increasing sense of separation anxiety. At least one can know the source of every problem, and have something and someone to blame—by turns humanity in general, Western civilization, and Americans in particular. Radical environmentalism has become the central organizing principle of a new cult, an oracle that provides all answers.

Not content simply to save the natives in some distant demesne, the followers of the Green Dragon find consummation at romantic locales like Rio and Bali, where they hatch ever more grandiose Green schemes. Al Gore is usually prominent on these occasions. Having conquered the political,[67] literary,[68] and movie[69] galaxies, it is only apt that his next ambition is to teach mankind its supreme purpose in life, and how to attain it. We must "save the planet." Ironically, *we* must save the planet from *us*, and *now*, or at least by the new deadline of 2015.[70]

In his acceptance speech for the Nobel Peace Prize Gore alleged, rather grandiloquently, "we have begun to wage war on the earth itself."[71] Because of this war "the earth has a fever. And the fever is rising."

In short, we must make "rescue of the global environment the central organizing principle of the world community." Finally, he requires immediate global action to save her, much of which must necessarily be executed by government fiat and compulsion: "We must abandon the conceit that individual, isolated, private actions are the answer. They can and do help. But they will not take us far enough without collective action. At the same time, we must ensure that in mobilizing globally, we do not invite the establishment of ideological conformity and a new lock-step 'ism.'" Saving the Earth requires much more than driving a hybrid vehicle and having a compost pile; according to Al Gore, such actions are not only conceited, but ultimately futile for "they will not take us far enough…."

This disturbing collectivist vision already explicitly mobilizes government muscle to forbid building power plants[72] and includes plans eventually to ban the internal combustion engine,[73] all because they produce carbon dioxide gas. Even the infamous communist collectives, in China and the old USSR, did not deny their benighted people, by order of government, the poor light bulb,[74] or forbid a hearty fireplace.[75] Yet Western environmentalists dream up precisely these kinds of insane utopian policies and manage to have their governments steamroller them into practice.

Believe it or not, the US government has banned the humble light bulb. First they came for the flush toilets, and we said nothing. Then they came for the light bulbs, and we flushed our grumbling to the side. Will they next come for the people? The Green Dragon is a leviathan-like temperance movement on steroids. From branch to root this environmental-*ism* huddles close to the primordial muck of tyranny, and there is nothing liberal about tyranny.

## Green Religion

Environmentalists claim that humans "disrupt the entire web of life in the oceans,"[76] engage in "rape of the environment,"[77] and routinely produce "ecological holocaust."[78] In 2006 Al Gore suggested that we are "altering the balance of energy between our planet and the rest of the universe."[79] And right on cue some obliging scientists with environmentalist credentials found the smoking gun. Apparently our ruthless astronomical probing of her dark matter may be shoving the universe over the edge towards extinction.[80] Yes, that universe; the entirety of space and time is going extinct because our brazen telescopic leering caused deep embarrassment and intense cosmic shame to what some environmentalists claim is, as feminist eco-theologian Sallie McFague put it in the title of a book, *The Body of God*.[81] To hear environmentalist storytellers tell it, rather than being victims of that evil that slithered from Pandora's box, humans *are* that phantasmal evil.

Most devoted environmentalists claim to be much too sophisticated to be seduced by ancient pagan nature worship. And to be sure the environmentalist movement is far too broad and nuanced to be allured by pagan religion alone. Nevertheless, in a palpable way the environmentalist Litany provides the foundation of an alternative religion or worldview for those who reject Western civilization. One might argue that radical environmentalism is becoming "the central organizing principle of the world community."[82] It is a religion with a vision of sin and repentance, heaven and hell. It even has a special vocabulary, with words like "sustainability" and "carbon neutral." Its communion is organic food. Its sacraments are sex, abortion, and when all else fails, sterilization. Its saints are Al Gore and the IPCC.

Guilt ridden? Well, now people with money to burn can buy indulgences just as in the medieval church. In that religion priests sold indulgences to souls burdened with guilt over their sins. The Green movement has absorbed these faux religious elements, and for each one has its own ersatz affectation. Whatever it may mean, raping the Earth is not ecologically sound or morally attractive. But if you must rape Mother Earth, then be sure to visit the Green Dragon and pay a sum of money[83] to cleanse your guilt, until next time. Forgiveness for sins is only a carbon offset away.

It is possible to calculate the extent of one's sins online. According to TerraPass, over the past three years my weekly driving has resulted in about 5,224 pounds of $CO_2$ a year, and for a mere $29.95 I can buy an indulgence that will offset the environmental impact of my reckless, indeed sacrilegious, lifestyle. These carbon offsets will do as little for the salvation of the world as indulgences would for my soul. But for people with a desperate spiritual hunger they are a panacea and penance they are eager to embrace, and an incredible gift to give—a gift that promises guilt-free living and a purpose-driven life. If only it were true. Go ahead, say the Gore-like business barons, pay the toll at our gate, and this year "Give the gift of green."[84]

This is just among the first tender strokes emanating from the Green Dragon, one of a long list of required sacrifices and penances. In his book *Green Rage*, Christopher Manes enumerates some more, including elimination of all use of oil[85] and the deliberate quest for poverty through deindustrialization.[86] These prescriptions will lead the plain people to stagger under the self-flagellating burdens of environmentalism, its endless obligations and corollaries.

Massive sacrifices become a moral imperative since human beings "are actually desecrating the earth."[87] So the Earth, Mother Nature, Gaia if you will, is sacred. But humans are a cancer, and a virus. There is a small grain of truth in this statement, for we read in the beginning of the Bible that the first desecration of the Earth occurred by the hand of man, when Adam sinned by eating the forbidden fruit. This intensely simple act has led to all sadness and horror and misery on Earth. Before this terrible sin and fall, Earth was spotless, but it became horribly changed so that wherever the man might look, the atrocity of his sins would meet his eyes. Man's disobedience led to God's cursing the Earth (Genesis 3:17). His anger, like a flood, overflowed all of its parts.

Not long after Adam, godless Cain murdered his brother, whose life was, as it were, consumed by the Earth that "opened its mouth to receive [his] brother's blood" (Genesis 4:11). In abhorrence of the pollution Earth, personified, opened her mouth to cover the blood. It was like there was, according to John Calvin, "more humanity in the earth than in man himself."[88] But the pollution was not man's blood *per se*. Man's blood was too precious to lie exposed, because the human blood mediates the precious human life. Human beings are not garbage on the Earth, not pollution, not cancer, not even a virus. Despite the horror inflicted through Adam's sin and propagated through shared blood that flows in the veins of over 6 billion humans living today, this same blood shares something with Jesus too, and so the blood is not without redeeming qualities.

## Useless Breathers

But 6 billion is a lot of people—6,000,000,000 souls. Grumpiness about more and more humans violating Mother Earth has become *haute gestalt* and finds its voice in the radical Green movement. If one should dread increasing population as the scourge of our species, as Malthus claimed, and if one should not allow freedom for the political instincts of the plain but ignorant folks, as Al Gore proposes, then what is the solution?

Freedom for the plain but ignorant folks only serves to increase the population, so personal liberty must decrease, and the population must decline—specifically, the liberties and numbers of the plain folk. The Voluntary Human Extinction Movement[89] is one example of organizations that hope to play an important role in furthering these goals. Its manifesto states: "Phasing out the human race by voluntarily ceasing to breed will allow Earth's biosphere to return to good health. Crowded conditions and resource shortages will improve as we become less dense." Dense indeed. These folks may be the alleged sharp edge of environmentalism, but the same ethos is also expressed by green celebrities who surely represent the dull and fuzzy parts. They believe in evolution and so, at least in theory, understand the value of species survival, but they have decided it is not for us.

In Western Christian civilization, married women have traditionally enjoyed a monopoly on the pleasures of sex and babies, at least theoretically. For this reason, those who never married were often pitied. But not a few people think celibacy is somehow ennobling (which is questionable in light of Genesis 1:28; 2:18; 9:1; 1 Corinthians 7:28; 1 Timothy 4:3; and Hebrews 13:4). Malthus considered the stigma associated with what he called "old maids" to be unjust prejudice, and the praise and honor associated with motherhood to be ill-earned gains. He viewed the mother of many as a self-indulgent monopolist of scarce resources, and the unmarried woman a selfless benefactor of society, noble because she, like the nun, chooses not to have children. In his 1798 *Essay on the Principle of Population*, he wrote, "Like the truly benevolent man in an irremediable scarcity, she has diminished her own consumption, instead of raising up a few particular people, by pressing down the rest. On a fairer comparison, therefore, she seems to have a better founded claim to the gratitude of society than the matron."

Poverty as the certain rotten fruit of an excessive population implied that those who chose to have children willfully contributed to scarcity. Their bourgeois and common children ate the food that others needed. In contrast, those who chose to forgo the immediate pleasures of sex and children would add to the happiness of society due to increased stores. Malthus and his kin are mistaken, for they have chosen to ignore the evidence that human poverty is usually a result of inferior cultures, or excessive government, never ultimately of excessive population. Models of bacteria in a Petri dish simply are not analogous to human civilization.

Devout disciples of Malthus have no room for sex that bears fruit in children, fulfilling the mandate of Genesis 1:28. To prevent overpopulation, they prefer sex without offspring. True believers like Toni Vernelli would fit right in. She cares so deeply for trees and animals that she has denied herself children for the sake of the planet. Ten years ago she aborted her first child. It was not that she was too busy in her job coordinating the European ops for PETA, the warm and snuggly animal-rights group. She just "passionately wanted to save the planet—not produce a new life which would only add to the problem."[90] Eight years ago she ended the life of her second child by abortion, before finally managing to hunt down a doctor that would agree to sterilize her. She now enjoys a satisfying vegan lifestyle, recycles, claims to have a small carbon footprint, and no (living) children. Her only frustration is the occasional person who does not appreciate her impregnable perspective: "When I tell people why I don't want children, they look at me as if I was planning to commit murder."[91]

One way environmentalists hope to encourage emulation is to have a carbon tax on the folk that choose to have more babies than the almighty government decrees. Perhaps echoing calls for similar action decades ago from Paul Ehrlich, modern Australian environmentalists urge their government to enforce on the selfish breeders a $5000 fine per child above the decreed maximum of two, plus a yearly carbon penance near $800 per child.[92] Malthus would be proud. We do not yet throw newborns to the crocodiles, as did the ancient Egyptians on one occasion, but instead sacrifice them to a creed. Western environmentalists yearn after the Chinese one-child policy, where second or subsequent pregnancies receive an official government reward of fines, forced abortion, and sterilization. Malthusian Greens tout this as selfless, sacrificial, ennobling, and deeply moral, but it is merely a reversion to totalitarian salvation by force and violence.

## Conclusion

Now that moral relativism so permeates society, including cultural organs such as government and the educational establishment, should it be a surprise that the foundations on which our ancestors built the wealthiest, most advanced civilization in history are under radical attack? Christianity eschews relativism and presents man as a moral and rational creature, separate from animals, and capable of knowing and understanding truth. To reject this view, upon which Western Civilization was erected, and to embrace relativism is to embrace irrational anthrophobia, fear of mankind. Contrary to the view of the founder of animal rights group PETA, it is not true that "A rat is a pig is a dog is a boy."[93] We humans are special creatures, in a class of our own, quite separate from, and superior to, trees and animals.

Radical environmentalists understand the nature of their unholy mission. To make the environment the central organizing principle for society is contrary to the Christian view. Peter Singer, Princeton University ethics professor and author of the cult classic *Animal Liberation*, has argued that "Judeo-Christian religious tradition is our foe."[94] This is because Christianity is not a biocentric religion and thus is opposed to idolatrous Green imperatives. Confusion lies along this slimy jade road. As we will see, it is paved with all kinds of perverted and destructive behaviors, leads to death itself, and finally, to the pains of hell forever. Instead of meat, hardcore environmentalists eat the moral reserves of previous generations and seek to prevent the production of new reserves and resources. The Green cloak that wraps the self-infatuated and self-destructive beast drips hatred of man.

Past decades have witnessed the accelerating ascent of what one could call, in its ugliest manifestation, the environmentalist death cult. It appears environmentalists believe that smart people have fewer children,[95] and the wisest have none at all. From the writings and public declarations of prominent environmentalists it is clear the ethos of the movement fosters deep animus and hostility towards humanity. They hate humans. They hate them for being fruitful and multiplying.

They hate them for having dominion over the other creatures, and for modifying nature. They reject the idea of humanity's special place in the world, and that God made man in his image. They spew disapproval of modern civilization. As we will see, they worship the wild wasteland and idolize nature.

Yet ironically the strength of the environmentalist movement lives in man; little man who is but broken human clay, soon to pass on, like fading mist on a warming day. The environmentalist movement is like a great dragon coiled over on itself, consuming its own tail. Sated with human blood, its self-loathing character and self-contradictory existence will end via its own delicate appetite. But as it roars and terrifies the faint of heart, it will also thrash about in its existential rage. It has done great harm to civilization, and is doing much more real harm to man. So it is imperative to hasten the death of the dragon, to limit the damage it does in its death throes, and to hope and work for something better.

The Green Dragon must die. As it is the symbol of a profoundly spiritual religious movement, the most effective weapons against it are those that are spiritual. It is high time; no, it is long past time the Christian view of the world again enjoyed public expression. If it is true that Biblical teachings on stewardship of the Earth do not enjoy vigorous promotion, then this is no excuse to become befuddled by the noxious Green odors and doctrines emanating from the foul beast. Unfortunately many professing Christians have become just so befuddled, and in some cases find more satisfaction in environmental activism than in Christ.

One cannot ignore certain attractive aspects of the movement, such as the meritorious idea that humans ought to ponder carefully the consequences of their actions on other creatures. But we do not need the environmentalist movement to point out these common-sense Christian values. Buying into the environmentalist ethos because a group opposes cruelty to animals, for instance, is like thinking it safe to pet a rattlesnake because its scales sparkle in the sunlight, or because it winked and smiled.

When the light exposes the unholy religious foundations of environmentalism, like the bones of a rotting carcass, no more disguise will be possible. No painted face, no fluffy illusions, no amount of bright semantic gymnastics, will morph the Green Dragon into anything other than what it is—an avowed enemy of humanity and of the greatest lover of humanity. No Hollywood celebrity bunnies draped over its foul form can deny its native evil. No precious perfumes can mask its vile vapors, its dark cloud of pantheistic pollution. We will not mourn its passing.

Compromise is not necessary. Since the environmentalist movement sets itself from the outset against God, it is anti-God and antihuman, exalting the Earth and almost deifying it as Our Mother Earth. In practice it razes man and tries to raise the Earth. How alien then, to the environmentalist thinking, is the Christian view of dominion and stewardship of the Earth! Such a view neither exalts nor denigrates the role of the Earth but provides a proper balance and perspective; the Earth is a creation of God, and He commands man to exercise dominion over it, to discover its secrets and benefit from its storehouses of treasures. Man is to recognize the beauty of the Earth and to care for it as a precious resource to put to wise use.

Yet the Earth, unlike man, will not continue forever. Flames of fire are its destination. The beast rages and refuses this fate, but it is futile to rage against God. No one, no man, creature, or idea, can stand forever in opposition to the will of God. In the end all must bow down before Jesus Christ and kiss Him, or perish in His wrath (Psalm 2:10–12). Unconditional surrender to the King of kings is the only reasonable choice. The Green Dragon will die.

# CHAPTER TWO

## The Church Complicit: How Environmentalism Has Penetrated the Church with Anti-Human and Anti-Christian Ideas

*Now I begin to be a disciple. I care for nothing, of visible or invisible things, so that I may but win Christ. Let fire and the cross, let the companies of wild beasts, let breaking of bones and tearing of limbs, let the grinding of the whole body, and all the malice of the devil, come upon me; be it so, only may I win Christ Jesus!*

Ignatius, Presbyter of Antioch (b. 35 or 50)[96]

## The Blame Game

Every strand of the political spectrum agrees that as the greatest civilization in human history unravels, like a ball of yarn strung out in time, it falls far short of its ideals. On the one hand many folk, including radical environmentalists, lament the lingering Christian character of Western civilization, for instance the view that economic growth and human progress are good.[97] They deny or reject the Christian roots of that civilization. But on the other hand, as far as it concerns them, they like the fruits, such as increased quality of life, and vainly imagine they can preserve the fruits after chopping the roots. They pine for a new age of zero or negative growth where the few humans that still exist to pollute the Earth will engage wholly in social justice, whatever that may mean to them. Christian influence they invariably characterize as blight on society, and its remaining residue as unsightly scarring.[i] The reality is otherwise, as described by Max Weber.[98] "The culture of the West," remarked historian Herbert Schlossberg, "infused as it is with Christian values, is superior to any other, and all the valid charges against the West are indications that it has betrayed its own heritage. It is not superior because it is wealthy; it is wealthy because it is superior, because it believes that work is a calling, that matter is important, that reason is a gift of God. This culture, God's gift, transmits its material blessings along with its interpretation of reality."[99] It is not that Muslims, Hindus, Buddhists, or even Atheists are somehow inferior; it is just that their ideologies are. In fact, when pagans adopt Christian virtues, more or less, as is the case of large dense swathes of Asia, blessings follow.

Since World War II wealth and human well-being have continued to progress in the Western world, but freedom and civilization have receded. Devaluation of human life, increase of violence against the weak and defenseless, and societal celebration of lust, lasciviousness, and lewdness are simply a few bitter fruits of the betrayal of our Christian heritage. Perhaps the Green Dragon is just one instance of a larger problem, a prominent sect in a meandering maze, one cultural

[i]Authors like Jared Diamond (*Guns, Germs, and Steel* [W. W. Norton & Co., 2005]) argue that Western civilization is not so much a product of ingenuity, but of luck and location. This should be contrasted with the excellent work, *Conquests and Cultures*, by Thomas Sowell (New York: Basic Books, 1999). Both reject the racial explanation for Western dominance, but where Sowell recognizes the influence of geography and climate, and gives to culture the central place, Diamond ironically sees mainly blind luck.

expression of a primal philosophy. If so, then before identifying more clearly the tangled roots of modern environmentalism, and doing a little chopping of our own, some introspection might be helpful. Who or what is to blame for this environment in which Western civilization is under such radical attack? The Green Dragon did not rise out of a vacuum.

Those that exercise the greatest societal influence are clearly prime suspects: government authorities, the media, educational establishments, corporations, and even fabulously wealthy individuals. These are the usual suspects. But what about the professing church of Christian believers? "Tell me what a person believes," said Ralph Waldo Emerson, "and I'll tell you what he'll do."[100] In the not-so-distant past the Christian church exercised great influence on public morals, and so influenced social and political policy. What John Cotton, the seventeenth-century New England Puritan minister, "preached on the Lord's Day was followed by the synod, and … what he preached in the Thursday Lecture was followed by the General Court."[101]

The Christian ethos undeniably shaped the culture and institutions of Western civilization; its fingerprints are unmistakable. In America, for example, where the Church sowed the Word of God new life sprang out of the ploughed ground along with innumerable treasures, such as Biblical respect for rational thought and learning. Thus arose great schools among which are Harvard, Princeton, and Yale.

Since all members of industrialized nations have passed through their various educational regimes, holding that knowledge is indispensable to power, perhaps the educational institutions are most responsible for the recession of civilization. After all, educational leaders herald their role in guiding development of society, in acting as change agents. Yet there is wide recognition that the modern American education establishment is largely[102] corrupt, often acting proudly in opposition to the intellect.

At Harvard College, formed just 16 years after the first Pilgrims landed in New England, the incoming student learned that "the main end of his life and studies is to know God and Jesus Christ which is eternal life (John 17:3) and therefore to lay Christ in the bottome, as the only foundation of all sound knowledge and learning."[103] But today the quest for knowledge and understanding receives active scorn on college campuses, and not chiefly by the students. Bright-eyed students have engaged with intellectually corrupt professors so in practice, "for many, a college education has meant an applied course in immorality."[104] Consider an example: In the twenty-first century, many colleges confound reason by practicing or advocating racially biased entrance requirements in the name of justice, or some similar moral inversion.[105] According to the influential 1983 government report *A Nation at Risk*, "Our society and its educational institutions seem to have lost sight of the basic purposes of schooling…."[106] One might argue that in the decades since the blindness has become profound.[107]

Through their mottoes most universities explicitly recognize their motivating principles. Finding them and dusting them off can come as an unexpected shock. Like stumbling over an abused puppy hidden away in a dark corner, rediscovery serves to stress the miserably backslidden state of the modern educational establishment. If one looks at great universities of the West it becomes interesting to note, though it should not be a surprise, that almost all their mottoes refer to or paraphrase or quote directly from the Holy Bible. Oxford University has the Latin motto "*Dominus illuminatio mea*" from the opening words to Psalm 27, translated "the Lord is my light." Harvard University's original charge was "*Veritas pro Christo et ecclesia*" (Truth for Christ and his Church).[108] For Princeton University "*Dei sub numine viget*" (Under the Protection of God She Flourishes). Dartmouth College claims "*Vox clamantis in deserto*" (The voice crying in the wilderness) from Mark 1:3. Yale University points to 1 Samuel 14:41 in its "*Lux et veritas*" (Light and truth). Columbia University quotes Psalm 36:9: "*In lumine Tuo videbimus lumen*" (In Thy light shall we see the light). Johns Hopkins claims "*Veritas vos liberabit*" (The truth shall make you free) from John 8:32. One could go on and on like this for some time without any difficulty.

Today many, perhaps most, universities of the West have aggressively rejected the Christian vision articulated by their mottoes, and the idea of Christian scholarship has become an outrageous oxymoron. Degradation of the pursuit of truth is not to the same extent in all universities, nor equal in all departments of the university, but is part of the milieu of academia as a whole. Schools retain the mottoes as a façade, perhaps out of nostalgia, perhaps for other reasons. But they have, if their mottoes are any suggestion of the way things once were, wandered far from their erstwhile principles and ambitions.

## Just Words

It seems almost a copout to blame universities—their many impious professors and spineless bureaucrats—for the sorry state of affairs, and it would be. Intellectuals certainly bear great blame for magnification of the worst, and degeneration and dissipation of the best, in Western civilization. But the greatest blame we must lay squarely at the foot of the professing Christian church. Her members are to be representatives and ambassadors for the better way, the way in opposition to the pagan emphasis on the sensual so common today. Yet many have idly watched or actively aided societal decay. Either way, the church is complicit.

In the Bible, cultures and civilizations are spoken of not as rising or falling like a cloud or a rock, but as being under the curse or blessing of the almighty God. That a cloud cannot rise, or a sparrow fall, without His oversight should make clear that even the nations are like so much clay in a potter's hands.[109] A radical environmentalist goal for society, namely stasis or sustainability, ignores God's hand in history and pretends that society is amoral. But the Bible teaches that God has an interest in the behavior of humans, and, "It is a fearful thing to fall into the hands of the living God" (Hebrews 10:31). We may be grateful that God is a merciful Sovereign, and slow to anger, for if He curses a society, no amount of personal or central planning can save it from disaster. History is therefore not a matter of fate, or man's abilities, but of the decree of God. Any civilization receives blessing insofar as the residue of Christian values is appreciated and nurtured, but cursing and degradation when it is despised and ignored.

Jesus said the quality of society relates directly to the degree to which Christians are faithful to follow after His teachings revealed in the Bible. He said of Christians (Matthew 5:13), "You are the salt of the earth; but if the salt loses its flavor, how shall it be seasoned? It is then good for nothing but to be thrown out and trampled underfoot by men." So if a society is overflowing with the sewage of wickedness and injustice, the Church bears some responsibility. For how can one expect any society to have a healthy savor and light without the peaceful influence of Christians living out their faith?

Therefore, when Christians set anything, or anyone, above or before Jesus Christ and His gospel, they are good for nothing. They may do good deeds before the eyes of men, but if they love not what Christ loves, they are good for nothing. Important as good works or social activism may be, they can never replace, displace, or substitute for the dominating focus of Christianity, articulated by the Apostle Paul: "I determined not to know anything among you except Jesus Christ and Him crucified" (1 Corinthians 2:2). While most professing Christians would agree with Paul's words, they might not agree with Paul's meaning. A classic example is the young man who insists his professions of love entitle him to carnal knowledge of his girlfriend. He has redefined love to mean satisfaction of his lusts, but his friend, holding to the historical definition, may foolishly succumb because she naïvely imagines he cares for her well-being.

To communicate rationally, words need content and precise definitions. Unhappily, modern Christendom has shown a decided turn for the irrational, and professing Christians seem often to prefer a studied vagueness over Scriptural clarity. Followers of Christ desire to submit to the concrete, explicit teachings of the Bible; this is their mark. Vague principles are not enough; the actual *words* of God must be followed, and not wrested out of context.[110] Anyone can wrest Scripture. For example, swami Satchidananda, who was the opening act at Woodstock (1969), taught in San Francisco that when Jesus said, "Blessed are the pure in heart: for they shall see God," he *meant* blessed are those who upgrade their consciousness, because they shall recognize that they are themselves God.[111]

Satan also twisted Scripture, but Jesus rebuked him and pointed out again and again that authority is derived from God's Word alone: "It is written," said Jesus "'Man shall not live by bread alone, but by every word that proceeds from the mouth of God'" (Matthew 4:4). What a scandal it is that so many professing Christians find words like these passé! Since they have adopted the spirit of this age, they think it fine and dandy to redefine words and their context like a rubber band. When professing Christians say they agree to Scripture but stretch its words to justify a secular substitute, they are acting no differently than the swami. Except as a word to exploit, Christ is no longer precious to them and becomes instead a stone of stumbling and a rock of offence (1 Peter 2:6–8).

Having turned their eyes from Christ, seeing Him as of little practical use, many professing Christians are become empty of value and seek to be filled otherwise. And filled they will be. They and their society suffer the curse of God. Former U.S. Vice President Al Gore is a prime example. Today his declared central organizing life-principle is salvation of the environment. Yet for most of his life he was a member in good standing of the 16-million-member Southern Baptist Convention. In 2002 Mr. Gore told *Newsweek*: "I am a Christian. I am a Protestant. I am a Baptist."[112] But by 2005 he was no longer a member, having moved on to a more compliant form of Christianity found on a different rack at the spiritual supermarket. "We're ecumenical now," Gore mused pensively, maybe with even a hint of sadness, to a reporter. In the same interview his wife said, "I think I follow Baba Ram Dass," [113] that eminently ecumenical psychedelic Hassidic Hindu.

Gore's long and winding road rushing away from the church was obvious years before, and is sure to eventually bring scandal, shipwreck, and sadness to his life. In his 1992 book *Earth in the Balance* he came to the position, long taught by others, that men should harness religion as a political tool. "The fate of mankind," said he, "as well as of religion, depends upon the emergence of a new faith in the future. Armed with such a faith, we might find it possible to resanctify the earth."[114]

Such cold and formless religiosity is no substitute for the true Christian religion, which is manly and robust. Gore also lamented the jarring role of traditional Christianity in displacing Earth-focused consciousness and pointed favorably to pagan pantheism. For example: "The spiritual sense of our place in nature … can be traced to the origins of human civilization. A growing number of anthropologists and archeo-mythologists … argue that the prevailing ideology of belief in prehistoric Europe and much of the world was based on the worship of a single earth goddess, who was assumed to be the fount of all life and who radiated harmony among all living things."[115] To this day he contends this ancient Gaia-centered faith is the common thread that raises all flags, that in some way directs and allows us to discover the commonalities in the religions of the world.

And so Gore calls for a new faith—or more accurately, a redefinition or reinterpretation of one's personal faith tradition that recognizes this original pure pagan thread. He wants a faith that finds the image of God in the creation, a faith that will sanctify all matter and set it apart as holy ground. This is nothing less than a naked attempt to transfer sovereignty from the Creator to the creation, to transfer autonomy from God to man. The reality is that God alone defines the holy ground. Gore's arguments for a "panreligious perspective"[116] simply do not conform to the old and unambiguous Biblical absolutes and are, in fact, much more in keeping with the Hindu conception of *shakti*, the dynamic divine feminine force.

Gore is not alone in sloughing off the historic Christian view of the world. Under the guise of ecumenism and environmentalism he and a host of professing Christians wholeheartedly embrace mixing and matching of the best traditions that religion has to offer. In this view Christianity is a tolerably harmless ancient delusion that wants an upgrade, like the change from one computer operating system to another. Indeed, Christianity may even be handy if mainline American denominations can muster their historic moral authority, still vaguely remembered in the general culture, to engineer a new religion of power. This will do for the churches what a logic bomb does in the guts of a computer. Stanley Fish, the rogue postmodern and no friend of Christianity, recognizes that the historical record shows true Christian faith changes everything else. In his well-read essay *Why We Can't All Just Get Along*, Fish is taken aback that professing Christian leaders have been willing, even eager, to embrace causes antithetical to Christianity.[117]

It is not that Christians do not share common goals with radical environmentalists, such as the wish for less pollution or prevention of cruelty to animals, but that they should recognize and resist the antichristian presuppositions of the environmentalist movement. Stalin justified the murder of tens of millions through his social policies, during peacetime, but he also helped defeat the Nazi menace. This benefited the Allies but did not suggest any obligation for them to surrender to communism. Or to put it differently, the pretended innocence of those previously found to be dangerous does not trick the wise. Neither should we allow the Green Dragon to woo us.

Many intellectuals and politicians are only too happy to exploit Christianity, indeed any other religion, all and any, as a means to achieve their goals. They seek eagerly for new ways to recast their particular visions of salvation—eco-salvation or otherwise—in terms that will attract naïve religious voters, like a moth to a flame. For example, in 2005 the *Washington Post* noted, "After the election last fall, leaders of the country's major environmental groups spent an entire day at a meeting in Washington trying to figure out how to talk to evangelicals."[118] Politicians pay attention because in America 78 percent of the population claims to be Christian, and over 90 percent share a belief in God.[119] According to reliable media sources, Bible-believing Christians are "largely poor, uneducated, and easy to command."[120] They are the perfect political constituency, ripe for shearing, or worse, by wolves in sheep's clothing.[121]

## John Dewey's Silly Sheep

Modern American education reflects the philosophy of educational theorist John Dewey more than any other. Dewey is a worthy representative of those who in the early twentieth century magnanimously enlisted churches to join the social justice crusade of the humanists, to remove enslaving ideas—for example belief in the one living and true God. Mainline churches have long since surrendered intellectual leadership and authority, having steadily pared every Biblical doctrine to the point that they whittled down Christianity to mere feelings.

Instead of having a sharp edge—the Bible teaches that true Christianity is an offense to those that are perishing—the mainline churches have become dull and without edges at all, a designer religion without any cutting power. They are cold like ice, smooth but slippery. Dewey, in the twentieth century, like others in the twenty-first, was perfectly content to make peace and common cause with a naturalized version of Christianity in which upgrade or redesign of central doctrines ensured no offense to the rationality and confidence of a modern secularist.[122]

Dewey perceived nervelessness in the Church for whom heaven was but an accessory to worldly desire, for whom the world was the all in all, the center of all fond thoughts. So he suggested a way for the Christian church finally to leave the field of battle with some pride intact. With some handy refinements it would not be a great leap, indeed a quantum leap, to upgrade to the dynamic idea of an ever-changing God that was "a unification of ideal values," and the "*active* relation between ideal and actual."[123] One could now redefine God vaguely as an idea, or an emotion. Much as a rubber band conveniently adapts to the form of the object it embraces, such a smorgasbord God can fit just about any ideology.

But God is not an idea or action. God is not a feeling or an emotion, not simply a malleable part of speech. An infinite personal God does exist, and this supreme Being has revealed Himself to man. Dewey mixes sweet wine with sour vinegar. But his vinegar will sooner sour the gospel wine than the wine will sweeten the vinegar.

Glorious, ecstatic, unspeakable feelings that fill one with awe and satisfaction do not discover truth; neither does pragmatism. Reliance on feelings is a dangerous, muddled notion, inimical to Biblical Christianity; it will leave a sour taste. Jesus defined truth by God's Word (John 17:17), not by a moving experience or passionate conviction bought on a mountaintop or in a movie theater, or anywhere else. Martin Luther (b. 1483) wrote: "You should not believe your conscience and your feelings more than the word which the Lord who receives sinners preaches to you."[124] Luther was simply following the example of early Christians. With their simple faith in God's truth, they reveled in His eternal words, and in so doing injected leaven into the social and political order and turned the world inside out. True religion, now and always, should take its form solely from the revealed will of God. Wobbly Christians everywhere need to remember the power of the gospel. It is not reduced in our age.

Yet leaders in large swathes of the professing church, whose mainline denominations have progressively rejected Biblical authority for a feeling- or action-oriented faith, are drunken with the intellectual tsunamis of the world, which they like to think are the Green wave of the future. They would rather run with the wolf than with the sheep, rather fly at giddy heights with the eagle than with the lowly dove. In many cases, doubtless, the motives behind Christian environmental and other social action are good, but separated from sound Biblical understanding the potential for deception is great.

To be sure, Christians must *feel*, if they are Christians; Christians are not Stoics. And they must *act*, as the Apostle James so clearly teaches. That said, it is hard honestly to accuse modern Christians of being hard-boiled, ivory-tower intellectuals who only occasionally dip a toe into the world of men, that is, the world of emotion and activism. Rather, the frequent accusation, with some good evidence, is that they have soft heads. The problem is that their focus is so easily distracted from where it should be, on Jesus Christ. It is not that they think too much, but that they think too little, or not at all. Christians are, if they are anything, disciples of Christ. This means that their primary work is not to seek to "change the world" through political activism. As ambassadors of their heavenly Father, their primary work and consuming focus is to proclaim the written words of God, His merciful message to mankind. If Christians do this, the world will change.

## With the Force

One reason that modern Christians are so touchingly naïve about the dangers of environmentalism is that philosophies and teachings of the world have softened them up. Danish theologian Søren Kierkegaard (b. 1813) is one whose philosophy permeates much of modern Christendom. Kierkegaard argued that religious progress and illumination come through an irrational leap of faith, not through the logic of the Bible.[125] Christian philosopher Gordon Clark stated of Kierkegaard, "The fatal flaw is his rejection of logic. When once a man commits himself to contradictions, his language, and therefore his recommendations to other people, become meaningless."[126]

*What* a person believes is thus not important at all; what matters is *how* he believes. So a Hindu can meet God in earnest prayer, an atheist can meet God on a mountaintop acid trip or through passionate social activism, and a Christian can meet God by having a personal relationship with Jesus; boiled down it is all the same. The extent to which this rejection of intellect sounds reasonable to the professing Christian marks how deeply non-Christian thought forms have penetrated the Church.

It is this unscriptural focus on the how, not the what, that makes modern Christians susceptible to environmentalist mumbo jumbo. That is why many professing Christians applauded when Al Gore, sounding a little bit like a Jedi master in his Nobel Peace Prize lecture, said, "When we unite for a moral purpose that is manifestly good and true, the spiritual energy unleashed can transform us."[127] They say if humans can only channel this spiritual *satyagraha*, the pagan "truth force" that Gore is so fond of, then man would finally have incarnated the ubiquitous power of the universe. The force would then be with us, but would Christ?

## Greener Pastures

A Christian is one who wants to find out and do what pleases the Lord Jesus (Ephesians 5:10). One can therefore distinguish a Christian movement, in principle, by the willing-hearted and ungrudging resolve to submit to the demands of Scripture. "You search the Scriptures," said Jesus (John 5:39), "for in them you think you have eternal life; and these are they which testify of Me." Dishonest people may use similar Biblical language, but with new and unbiblical meanings. Since so much of the Church has exalted feeling or action over the Word of God, it should come as no surprise that significant parts of the professing Christian church are in the vanguard of the environmentalist movement. Rather than being fools for Christ they are useful idiots of the Green Dragon, well-meaning and infatuated sycophants, or cunning and sardonic hypocrites.

In the former case it would be wrong to suggest that religiously motivated environmentalists are merely crass manipulators grasping for the Christian brand. However, in the latter case the Christian label is sought as an umbrella of shelter against accusations of hidden motives, a source of influence and power, and a means to advance their political program; sheep shearing advances apace. Thus it is common to see advertisements of Jesus as the Marxist, organic-vegetable-eating, hemp-wearing, pro-choice, Prius-driving, flag-hating hippie. When Christianity is merely an ideal, Jesus and His narrow Way become nothing more than different strokes for different folks. Of course, false religion is a double-edged sword: it is equally dangerous to claim Jesus as a Neo-con, flesh-eating, flag-waving, gun-toting, truck-driving, gay-hating redneck.

Despite the seeming antinomies, the real religious differences between the Left and the Right in America are remarkably small. A 2007 Barna Group study found that 48% of Democrats are "absolutely committed to Christianity," and 41% attend church at least once a week. For Republicans the numbers were 61% and 53%, respectively.[128] The most significant differences found were in beliefs rather than behavior. A 2008 Pew Report confirms that only 14% of professing Christians cite their faith as a major influence on their political views.[129] This shows that most professing Christians feel that what they believe is a separate compartment of their life. What they believe may make them feel warm and fuzzy on Sunday but does not in great measure guide what they do or how they live. Or worse, the polls may suggest that they do not believe the Bible at all. They have been won over to *some* kind of belief, they are very religious, but the question is whether they have been won over to Christianity. The Bible suggests a dangerous and disturbing possibility: "But be doers of the word, and not hearers only, deceiving yourselves" (James 1:22).

## Fat Free Religion

The religion of the Bible, the Christian religion, is edgy and demanding. It is a religion of power, yet not a power religion. But Christianity is much more than a religion, it is an ever-regenerating power; history has displayed that transformational power of God through the preaching of the gospel of Jesus Christ. Thus a major problem with the Green Dragon is not so much that it is a revolutionary movement, but that environmentalism is insufficiently revolutionary. That a hypocrite like Al Gore is an environmental champion, that chorus girls like the Dixie Chicks can score points by having low-carbon music tours, shows further evidence for the spiritual shallowness of the movement. Environmentalism is eco-chic for hippie chicks. Surely the Church should not obsess over carb-free substitutes when it can have that which is sweeter than honey.

Religious indifference, moral relativism, and savage nihilism fill the atmosphere. In such an increasingly impoverished cultural climate, the influence of the churches for good is so corrupted because so many of the so-called church shepherds have replaced the revolutionary message of Jesus with vain and flashy frauds, like the Green Dragon. Faced with the threat of annihilation, timorous church leaders have been prepared to accept the truce offered by men like John Dewey. The awesome power of modern science, now almost completely materialistic in philosophy, once seemed poised to sweep all before it. Though not so proud today, science has attained unparalleled muscle and authority. Power to kill, yes, but religion holds the beating heart of billions—it has more social power and more influence for changing behavior than has materialist science. What if they could be united to be part of the new Green religion?

Biologist and belligerent atheist E. O. Wilson would woo Christians to see the possibilities of such a union, a Grand Unified Theory of life. In his 2007 book, *Naturalist*, he identifies science and religion as "the two most powerful social forces in the world."[130] In this work Wilson toned down his trademark militancy, as if the work of Wilson the younger were youthful exuberance; now he is just baring his fangs, if not lunging for the throat. Wilson the elder tries to make nice because the obvious societal disintegration can unnerve even tough-guy atheists. He believes that society works better if man acts as if there is a god, a useful hangover of evolutionary psychology.[131] Thus religion can provide obligatory sociological forms and encouragements, while science can provide the power.

Wilson concludes that science and religion must work together to achieve salvation of man and nature. He suggests policies that can lead to what he calls a sustainable new world order. For example, Christians should stop following the teachings of Christ, and rather work to reduce the human population to acceptable levels. Wilson wants Christians to grow up and evolve, and compromise a little. In *The Creation* he writes, "What are we to do? Forget the differences, I say. Meet on common ground."[132]

Perhaps Wilson dares to make so bold because he recognizes (1) how different modern mainline Christianity is from the religion of Jesus, and (2) how so many modern Christians are susceptible to his arguments since they base their faith almost exclusively on mysticism and emotional experiences. Wilson is himself somewhat the mystic. A *Salon* magazine article quotes him as saying, "Spirituality in my mind is a sense of awe and wonder. A combined sense of awe and wonder and mystery that comes from the perception of, or belief in, something that exists in the world that is very important to us…."[133] Since hallucinogenic drugs can replicate those same so-called religious experiences and feelings, he concludes that the God hypothesis is not necessary.[134] Wilson's thinking reflexively echoes a chapter title in Timothy Leary's autobiography *Drugs are the Origin of Religion and Philosophy*.[135] In that case, if evolutionary pressures require some kind of God hypothesis, then why can we not all agree to a more convenient god than the historical God of the Bible?

When asked how this new evolving faith will emerge to resanctify the Earth, Wilson replies: "That's easy: through the secularization of traditional religions. Already most of them accept the idea of evolution. Already most of them accept that the human mind really does have a physical basis. And, furthermore, most of the Abrahamic religions of Islam and Christianity have shown themselves very prone to the conservation ethic when the information is made available to them. There's a very strong green movement in Christian sects at the present time. So that is part of the evolutionary process."[136]

Christians once scandalized the Roman Empire by their no-nonsense embrace of Christian exclusivity. They were immovable in their conviction that there is one true and living God. Wilson's indecent proposals would have scandalized them. Why don't they scandalize us? In our times, as often in history, the supposed moral guardians—the mainline churches—appear to be motivated more by social interests; they are attracted by what J. Gresham Machen (b. 1881), quoting J. S. Phillimore, called "courtly polygamies of the soul."[137] They are scandalized by the gospel. Far too many of these have forgotten the transforming nature of their spiritual mission, seduced instead by evolutionary temptations to power offered by important men like E. O. Wilson.

Gordon Clark argued that the lure of power religion, the desire to exercise control over others, even by doing to or for them what is considered 'good', may be the shiny hook that draws people into deeper idolatries. Regarding science, Clark wrote, "The world at large holds science in such high regard that some Christians have begun to question the value of preaching the gospel. They have begun to share in the idolatry of science. Since they are professing Christians, reared in evangelical homes, retaining an attachment to Biblical views, these theologian-philosophers have picked out Gen. 1:28, coined the phrase 'cultural mandate,' and so emphasized subduing the Earth for man's comfort, that the ministry has actually been downgraded. Some have gone so far as to suggest that the preaching of the gospel should cease until society has been reconstructed along socialistic lines and thus made ready to accept Christianity."[138]

In seeking to make God relevant to the contemporary culture, those claiming to be its standard bearers have all but abandoned the Christian message. Philosopher Max Oelschlaeger is hip and *trés chic*. He is with the program, and believes that churches can and should become environmentalist covens of sorts. Though far too demure to put it that way, he sees potential, writing "In spite of the diversity of religious belief, there is a common ground for caring for creation that can make organized religion a political turning point."[139] Such manipulation is the way of power religion, not spirit religion. Spirit religion says, "For to me, to live is Christ" (Philippians 1:21). Power religion inverts the Biblical model and seeks inspiration and salvation in other gods, for example in seeking life purpose in political activism. This is putting the cart before the horse, becoming entangled in the carnal religions of the world rather than having concern for living in a fashion that is Christlike. Church leaders who seek to hitch environmentalism to Christianity are dirty fleas riding God's dog.

At his Nobel Peace Prize acceptance speech Al Gore said, "We have everything we need to get started, save perhaps political will, but political will is a renewable resource. So let us renew it, and say together: 'We have a purpose. We are many. For this purpose we will rise, and we will act.'"[140] Here is expression of religious purpose in life through political interests and activism. What an anemic synthetic substitute for genuine godliness! This deficiency is why many political activists, like Gore, try to gain higher ground by couching their language in the moral and spiritual.

Theologian Francis Schaeffer speaks of connotation language,[141] in which men vaguely use "god" words rich in spiritual connotations and without a meaning rooted in Scripture. They give an impression of communication when, in fact, by sleight of tongue they elicit a particular emotional response to dazzle and manipulate the hearer.[142] This is like luring people towards a cliff with their favorite candy when the real goal is to kick them over the edge. One sees this all the time in television advertisements; here is an attractive woman draped over an automobile, the potential buyer receives sexual connotations and hopefully the urge to drop his wallet. Similarly environmentalists exploit religious language—justice, peace, love, and so on—with the hope of manipulating the religious response into political activism.

They wish the dynamism of religious devotion, for which many are willing to die, would give impetus to their visions of a new world order.

## Behind the Power (the Throne)

Possibly the most influential environmentalist in the world is the tenebrous Canadian Maurice Strong, whose paw prints blotch the modern Green movement.[143] It is not an exaggeration to say that he has defined a significant proportion of the modern environmentalist agenda. He is the *Uber-Geist* behind Earth Summits I and II, held in Stockholm and Rio, the maker of international politicians, in the carbon business with Al Gore, the shadow behind the United Nations Secretary General for most of the past decade, and in the nation-breaking business with George Soros and other multibillionaires.

Strong, with these and other powerful and important men, argues that implementation of the Green agenda requires the little people or—more to the point—their Lilliputian politicians be sufficiently motivated. Impatient with the tedious delays and bean-counting that democratic civil society exacts he describes, in a 2001 interview, how to achieve global sociopolitical change, for example, via manipulation of economic levers to cause pain and thus influence behavior. Another powerful force to exploit is religious passions, says he. "People respond to their self-interest, yes," said Strong who is also passionately religious.[144] "But they also respond to their deepest moral, ethical, and spiritual values. And that's why, at Rio, I presented a proposal to world leaders that they adopt a kind of Magna Charta for the earth, an Earth Charter .... a statement of moral, and ethical principles to guide the conduct of people and nations towards the earth and towards each other."[145] The Charter, a list of some men's commandments, traveled the Earth in the *Ark of Hope*, a contemptible imitation of the original *Ark of the Covenant* that housed the Ten Commandments God wrote in stone.

Religious terminology, symbols, and passions become abused as central weapons in the radical environmentalist arsenal, moral devices to whip up guilt and motivate others to pursue ends that suit the putative masters of the universe. Given that so many wealthy schemers softly play the environmentalist fiddle, one wonders, with late author Michael Crichton,[146] whether salvation of the environment is just a feint, and no more than a means to more vulgar political ends. What Chesterton called "the appetite of tyranny" is alive and well. God knows the true motives. With Maurice Strong, though tempting to ascribe his environmentalist activities to mere *realpolitik*, the quasi-memoir *Where on Earth are We Going*[147] suggests that his grand visions for a new biocratic world order are a sincere product of his own confused spiritual pilgrimage. But whether motivated by power lust or the hungry soul's cry for salvation, or both—whatever the case may be—the Green Dragon is the bastard of spiritual polygamy.

The Green Dragon is only too happy to plow with God's heifer, especially if she is willing. If it is true that much of the professing church has been so tamed and conformed to the world's mold, seeking after the same power religion of the world, then there should be much evidence. And if so there should be repentance and great concern among true Christians and members of these churches. The problem of the Green Dragon, which we are considering, is then not only "out there" but much among us.

## The Church Complicit

As Cardinal Pell, Archbishop of Sydney, has pointed out,[148] radical environmentalists are beyond expert at moralizing their own political agenda. They are experts at exploiting envy, guilt, misinformation, and fear to control political outcomes. Church leaders are really not needed, but are useful props to mute the opposing and revolutionary nature of true Christian gospel witness. And perhaps better, if they are acolytes, they will advocate for the Green Dragon.

In the image of God is man created, says the Bible. The traditional perception is that the Christian church, with its strong anthropocentric currents, opposes the environmentalist *telos*. As such, and given the waning influence of mainline churches, it should not surprise that mainline Christendom thinks an environmentalist plank might bolster the *raison d'être*, and even win over secular critics. Christian historian George Marsden argues that lines between Christian and non-Christian morality are becoming increasingly blurred, but not because unbelievers are embracing truth. He writes, ". . . one of the great ironies in Western cultural history of the past two centuries is that many of the most effective attacks on traditional Christianity have taken place in the name of what amounts to a secularized Christian morality."[149]

Historian Lynn White in the 1960s made the case that the traditional Christian worldview was responsible for the mistaken view of man's relation to nature. Christianity, thought he, could be secularized or it could move out of the way. "More science and more technology," wrote White, "are not going to get us out of the present ecologic crisis until we find a new religion, or rethink our old one."[150]

The foreboding words out of academia do not go unnoticed, like the big stick the dog spies in the hand of the passing traveler. To the stick one might add the carrot of Methodist environmentalist Bill McKibben. He argued, in *The End of Nature*, that White's reading of the Bible was too narrow, that nature is on the whole depicted positively in the Christian tradition, which unfortunately has a hopeless addiction to the destructive idea of "progress."[151] This is the same tack taken in *Earth in the Balance*, in which Al Gore recapitulated René Dubos' arguments[152] from the 1972 Stockholm Earth Summit: we need a new or renewed religious model.

Rather than face complete marginalization among the intelligentsia, the mainline churches have elected to follow the old but modernizing plan of John Dewey. They are "getting with the program," and therein, perhaps they think, lies their survival. Consider, for example, the National Council of Churches (NCC), an association of 35 American Christian-related groups with about 100,000 local congregations and 45,000,000 adherents.[153] Among its members are a wide variety of mainline churches whose membership, it should be mentioned, has been in precipitous decline for decades. Most adherents, over 70 percent, are concentrated in twelve church groupings. In terms of these adherents the dominant players in the NCC, all with numbers over 2 million, are the United Methodist Church, the National Baptist Convention (USA, Inc.), the Evangelical Lutheran Church in America, the Progressive National Baptist Convention, the African Methodist Episcopal Church, the Presbyterian Church (U.S.A.), and the Episcopal Church in the United States of America.

During the 1950s the NCC entertained mostly innocent appreciation of the beauty of God's creation through its Christian Ministry in the

National Parks.[154] But by 1961 the NCC, little over a decade old, was working the cusp of the burgeoning Green wave when it issued a proclamation decrying human fertility and overpopulation.[155] Theologians such as Joseph Sittler, then of the University of Chicago, exercised enormous influence with tropes such as Earth rape and eco-justice.[156]

Most NCC member churches are proud sponsors of the Religious Coalition for Reproductive Choice, founded in 1973 to ensure the continued availability in America of unlimited abortions of unborn babies. Support for abortion, so they claim, proves their passion for the Christian cause and is evidence that they seek to express a prophetic vision of a just world through using "the moral power of religious communities…."[157] They call support for abortion reproductive justice; sodomy, sexual justice; welfare state socialism, economic or social justice; environmentalism, eco-justice; and so on. In the new program justice becomes meaningless because it stands for anything. Well, at least apparently anything that will overturn Christian morality and ethics in favor of left-wing prescriptions.

NCC patronage of progressive political causes seemingly knows no bounds. In 1998, Rev. Joan Brown Campbell, NCC's General Secretary, urged that belief in the radical environmentalist movement's assertions about global warming should become a "litmus test for the faith community."[158] Belief that Jesus Christ was born of a virgin, was crucified, and rose from the dead, in contrast, are divisive issues one should not insist on in polite company. Campbell is currently on the board of the World Council of Religious Leaders, an international ecumenicist organization that works to further United Nations visions to create a community of world religions.

One of the prominent departments in the NCC is the EcoJustice Program, "designed to heal and defend creation, working to assure justice for all of creation and the human beings who live in it."[159] In 2002, following the lead of the Evangelical Environmental Network, the EcoJustice Program embraced the "What Would Jesus Drive?" campaign, aimed at making Americans feel guilty about driving cars. Ironically, if He had begun His mission in the twenty-first century, Jesus might arguably have driven a fifteen-seat van, called Moby, to accommodate his twelve carbon-dioxide emitting apostles. This is the minimalist approach—a diesel bus would be a distinct possibility.

By 2005, when the global warming Kyoto protocol came into force, the NCC had enlisted thousands of Christian congregations in a massive letter-writing campaign in favor of environmentalist causes. So it is that Christian congregations become co-opted to train parishioners that it is their *moral* duty to go Green with Christ, to form Green Congregations, or Eco-Congregations; for the hip and chic, Green is the new black. The NCC EcoJustice program recently reiterated, "The opportunity is before us, to set the example and be a light in this world—Greening our churches, our homes, and calling on our leaders to fight climate change."[160] However, it is a little hard to imagine how they can remain light to this world given that they and their political allies have banned the humble light bulb and rage against the exhaust fumes from candles, wood fires, and anything else that shines or breathes. In camps, in Sunday schools, and from the pulpits comes the new Gospel of Green—to save the Earth from humans. Guilt manipulators of the world, unite!

Little wonder that in the fall of 2008 *The Green Bible*, naturally printed with soy ink on recycled paper, was unveiled with support from the NCC and various environmentalist groups. Together with the Bible, the binding contains a series of essays on radical environmentalism. Carl Pope, executive director of the Sierra Club, wrote: "*The Green Bible* is a unique treasure for people who want to more fully live out the scriptural values of stewardship, love for their neighbors, and care for the underprivileged. It provides inspiration and valuable resources for contemplation, community building, and action on behalf of the planet." Richard Cizik, while still vice president for Governmental Affairs of the 30-million-member *National Association of Evangelicals* (NAE), found the green letter Bible to be "exactly what the Church needs at this critical time,"[161] because it will become a tool to highlight what mankind needs, namely, "urgent changes in values, lifestyles and public policies to avert disastrous changes in climate." Cizik needs reminding that poverty produces pollution, wealth mitigates pollution, and the Word of God is not a recent discovery of *homo postmodernus*. Notice that he does not say the Bible is what the unbelieving world going to hell needs at this critical time. No, the green letter Bible is what the church needs. To be sure, when Cizik speaks of what the church needs it is quite possible that he is not referring to the words of God but to the radical environmentalist essays wrapped together with the Bible.

Cizik is one of a new generation of professing Green Evangelicals. He came to believe global warming is the greatest crisis in history in a Damascus Road experience. He claims it happened at an Oxford conference focused on exploring methods to focus Christianity on environmentalism. The conference was sponsored by Calvin DeWitt's Au Sable Institute, whose agenda is to bring Christians into the environmentalist fold by morphing churches into "creation-awareness centers."[162] At the meeting Cizik became shocked and awed by the apocalyptic predictions of professing Christian scientists whom he admired. While speaking with one prominent scientist, the scales fell off his eyes and he had an experience "not unlike my conversion to Christ."[163] Such is the power of the Green Dragon that he became born again.

Perhaps one cannot fault the motives of folks like Cizik; unlike the mainstream environmentalists for whom humans are scum, Green Evangelicals argue that we should save the environment because unless we act collectively and immediately the end is nigh for mankind. Unfortunately, they adopt the narrative of the Green Dragon and espouse the most radical solutions to environmental changes that are not—despite the hysterics—a crisis, often not even a problem.[164] Unfortunately, Cizik is increasingly less an ambassador of Christ than of the Green Dragon. He went on to work for Ted Turner's United Nations Foundation, an organization devoted to saving the Earth through global population control, and then later on for George Soros's Open Society Institute.[165] Bleeding heart evangelicals have swallowed apocalyptic nightmares of the Green Dragon hook, line, and sinker.

As always, the goal is to try to deflect the energies of Christians into environmentalist causes. One should realize the truth of God's Word is being presented alongside a good deal of environmentalist pollution. Instead of submitting political preferences to the Biblical test of unerring wisdom, the danger that religiously motivated environmentalists succumb to is the urge to sift the Bible through the teeth of spurious Green metaphysical and moral directives.

## Going Green with God

Formed in 1892 by John Muir, the Sierra Club is the oldest and largest environmentalist organization in the United States of America and works closely with the NCC. The Club is forthright about its hope to exploit Christian fervor, as if it were some powerful stream of consciousness to drive the Green wheel. If anything the Sierra Club has only become more radical since the early days when its founder, with characteristic subtlety, called the Christian God "purely a manufactured article as any puppet of a half-penny theater."[166] In 2008 the Sierra Club presented a national report on the environmental engagement of communities of faith, called Faith in Action: Communities of Faith Bring Hope for the Planet. "We also recognize," states the report, "that lasting social change rarely takes place without the active engagement of communities of faith."[167] They humbly discount their efforts and insist that environmentalism found in modern churches is a "spiritually motivated grassroots" effort.

What intrigues is that major revenue for the NCC no longer comes from its grassroots member denominations, but from private foundations and activist groups. Funding sources to the NCC put its political actions in a whole new light, since he who pays the piper calls the tune. Member donations to the NCC dropped 40% between 2001 and 2005, from $2.9 million to $1.75 million. Various secular foundations and non-governmental organizations, including the Sierra Club, were kind enough to pick up the shortfall. In 2005 private groups contributed $2.9 million to the yearly budget, an increase of over 350% from 2001, and greater than the giving from all NCC member denominations.

Talk of grassroots activism rings hollow. The money trail makes it increasingly difficult to characterize NCC political and environmental activism as grassroots. Current funding for the NCC from external gifts significantly exceeds and continues to outpace giving from the people in the member churches they claim to represent.[168] If the NCC were a business what has happened might be called a hostile takeover with the NCC acting now as a front organization for whatever causes the private foundations favor.

New infusions of Green for the NCC have come from left-leaning philanthropies including Pew Charitable Trusts, Tides Foundation, Rockefeller Brothers Fund, Turner Foundation, W. Alton Jones Foundation, V. Kann Rasmussen Foundation, and New World Foundation. These foundations funnel funds to fuel environmental politics. Take, as one example, the V. Kann Rasmussen Foundation, which provided funding for a series of conferences at Harvard

University in the 1990s entitled Religions of the World and Ecology Project. The goal was to establish common ground with all religions to reharmonize life on Earth.[169] This is only the tip of the iceberg. Through its operations in South Carolina the Foundation dares guide government policy directly. There it provides "funding to the Department of Education to support a statewide environmental education coordinator." In addition it seeks to deploy environmentalist propaganda to the youngest children through "funding for First Steps, an early-childhood education program supported by the State of South Carolina."[170] Perhaps it should not be surprising that Johnny in the government school cannot read, but Johnny knows all about sexual, social, and environmental "justice."

One certain lesson to draw from the Green money trail is that reverend bureaucrats of organizations like the NCC do not so much represent their congregations, but rather the political activist groups and private foundations. They have enlisted the professing church to drive their political program of mass persuasion. Or, if in their activism the NCC does accurately represent the churches, then it is fair to say that those congregations are not so much Christian churches any more as political action committees.

## Greener than Thou

Support for the NCC from external groups began to pick up substantially with the formation of the National Religious Partnership for the Environment (NRPE), of which the NCC is a member. Other major partners of the NRPE are the U.S. Conference of Catholic Bishops, the Evangelical Environmental Network, and the Coalition on the Environment and Jewish Life.[171] NRPE has an awesome army of organizational links and is itself something of a *matryoshka*, the Russian nested doll, which is never what it appears at first sight. For it is also not grassroots but launched with funding from Left-leaning foundations, primarily the Pew Charitable Trust.[172]

When it began in 1993, NRPE had over $5 million in grants to accomplish the goal, according to executive director Paul Gorman, of utilizing churches to produce a "distinctly religious response to the crisis of environmental sustainability and social justice."[173] The Partnership has been able to spread environmentalist propaganda and teaching aids to almost every Jewish synagogue, to every Catholic parish, and to several hundred thousand Protestant congregations in America. The Au Sable Institute, mentioned previously, operates a similar program.[174] Through it Au Sable has infiltrated Christian higher education by careful placement of teachers and teaching materials on environmental activism in schools associated with the Council for Christian Colleges and Universities (CCCU). Many of these schools are conservative in politics or theology. What they teach there under the Au Sable banner, will surprise their supporters.[175] Not surprisingly, the Au Sable Institute and the NRPE share a subset of the same donor pool. Little by little wolves try to douse Christian resistance and lead sheep by troubled waters to accept the inevitability of a divine environmental movement.

Seminaries also are now routinely cranking out a new breed of Church leaders trained, with materials from groups like the NRPE, in suitable guilt manipulation tactics and trendy ecobabble. The Episcopal Cathedral of St. John the Divine in New York City has been something of the epicenter for these initiatives. St. John the Divine housed and nurtured The Gaia Institute and the Temple of Understanding, the latter directed by that almost ubiquitous United Nations creature Maurice Strong. The NRPE is the 1993 spawn of this fount, the Temple of Understanding. U.S. Senators Al Gore, Timothy Wirth, and James Jeffords helped line up funding and arranged Congressional breakfasts for NRPE leaders. Al Gore even delivered a sermon at the Cathedral of St. John the Divine, calling on the congregants to recognize that "God is not separate from the Earth."[176] In other words, worship of Gaia is not that different from worship of God. Later in that same year, now as Vice President, Gore presided over the NRPE's inaugural meeting, commenting to the *New York Times* that the NRPE "will trigger the beginning of grass roots activity in tens of thousands of religious congregations across the country."[177] Gore's confidence, though appearing prophetic, was no doubt related to his insider knowledge of the big money greasing the agenda.

Another NRPE initiative, this time covering ten years and funded at $16 million by those in control of the Leftist philanthropies, drew to an end in 2009. NRPE has largely attained its 1999 goals, namely, to "assure that the next generation of religious leaders in America advance care for God's creation as a central priority for organized religion."[178] Here again is Gore's "central organizing principle." As if historic Christianity did not value nature, and the only ones who do were those who buy into NRPE propaganda.

Christians need not sway to the rhythm of subtle but patronizing emotional manipulations found in NRPE literature. Yet, through poor theology, a false sense of guilt, or an inability to parse fallacies wrapped in connotation words, religious leaders in predominantly mainline churches are hooked by what amounts to idolatrous environmental radicalism. The Foundation monies were well spent, for within the decade environmental activism came to prominence in mainline churches. Seminaries and churches across America now widely use for training purposes NRPE and other environmentalist literature.

From its associations with heathen Temples and Institutes it should be obvious the NRPE has an agenda that is anything but Christian, and it is not as if it were hiding its agenda. Yet few churches notice, maybe they don't care, that the NRPE is the outgrowth of the work of worldly "theologians" for whom the Bible is relevant only insofar as they may mold it to environmentalism and communitarian social customs. In this folly church leaders behave like a drowning man catching at razors. They play at Christian faith.

## Terrible Twins

George Orwell's *Animal Farm* tells of the visionary pig Old Major, who had a dream that soon proved disastrous: "And now, comrades, I will tell you about my dream of last night. I cannot describe that dream to you. It was a dream of the earth as it will be when Man is forgotten."[179] Among the trustees of the pagan Temple of Understanding, which founded the NRPE, was the eloquent (and heterodox) Roman Catholic priest Thomas Berry, who like the pig also has an anti-human dystopian dream. In his own *Dream of the Earth*, Berry argues for contemporary ecotheology and deep ecology, both terms that warrant explanation because of their baleful influence on the professing church.

The term *ecology* was coined only relatively recently, in 1867, by the German zoologist Ernst Haeckel , a disciple of Darwinian evolution. It is the study of the relation between organisms and their environment. "Deep" ecology reaches down and rails against "shallow" ecology, which focuses on such petty concerns as limiting human pollution for a healthier humanosphere and efficient resource management.[180] Deep ecology inverts the inherent anthropocentrism found in the previous definition of ecology, for deep ecology is the relationship between the environment and *her* organisms. The environment is subject, not object. It recasts ecology as a religious, not a scientific, pursuit.

What this means—this ecocentrism—is a rejection of the position that some life-forms have greater value than other life-forms. It means, as Aldo Leopold purred in his *Sand County Almanac* of 1949,[181] that human concerns do not matter much at all. Or to put it again in the words of John Muir (b. 1838), founder of the Sierra Club: "Nature's object in making animals and plants might possibly be first of all the happiness of each one of them, not the creation of all for the happiness of one. Why ought man to value himself as more than an infinitely small composing unit of the one great unit of creation?"[182] Ecocentrism thus becomes the central organizing principle that aims to topple mankind from their unjust and tyrannical perch as the crown of creation, the very image of God. Ecocentrism erases any meaningful distinction between man and nature, for all life is fundamentally one.[183]

Modern environmentalism and Darwin's dangerous idea are coupled like a nest of writhing snakes. If we take seriously the idea that man is an animal, it is not long before we understand, as Christopher Manes wrote, "evolution means that there is no basis for seeing humans as more advanced or developed than any other species."[184] Jesuit Teilhard de Chardin (b. 1881) was a religious popularizer of the view that not only man, but the whole cosmos evolves, for it, too, is a living organism.[185] Father Thomas Berry further extended the evolutionary link to a religious, moral, and ethical level. For Berry we must "resanctify" the story of cosmic evolution since just as man has evolved from the cosmic nothing so must his spirituality evolve.

## Ghostly Voices

Berry urges a return to the proto-ecology expressed by German romantics like Ernst Arndt. Arndt's 1815 paper *On the Care and Conservation of Forests* contains parts that give one the impression that modern environmentalists are really not so modern, after all. Arndt wrote, "When one sees nature in a necessary connectedness and interrelationship, then all things are equally important–shrub, worm, plant, human, stone, nothing first or last, but all one single unity."[186] In the extent to which mainstream environmentalists have accepted the razing of the distinction between man and other living creatures one observes a measure of the powerful influence of deep ecology. But far worse is the extent to which the church accepts this.

As just one example, consider the following from the Presbyterian Church (USA), in a document that claims the Biblical view recognizes "the impossibility of separating the human creature from the rest of creation."[187] The Biblical view is quite different and holds that God created the heavens and Earth distinct from Himself. Not only this, but there are clear distinctions within creation. It is not only possible to separate humans from trees, or animals; the Bible teaches that it is necessary. Of all God's creatures He makes man alone in His image.

In postmodern times the Norwegian philosopher Arne Næss (1912—2009), more than anyone else, became the still inner voice for a deep ecology. He redeveloped and popularized the doctrine of ecospheric classlessness—every part of the living environment, and the environment as a whole, has rights, and the same right to live and flourish. In this we see repeated the ancient heathen idea of a world soul of the living mother, Gaia, a fount from which all partake. Ironically, to ascribe feminine qualities to inanimate matter is hopelessly anthropocentric. But never mind, logic and rationality are not strong suits in deep ecology. Nevertheless, Western civilization must still be undressed, if you like, and liberated from the anthropocentric disposition of its Semitic roots.

Rejection of man's central place in the cosmos is a perennial theme in mainstream twentieth-century philosophy in the works of, for example, philosophers Martin Heidegger (1889—1976), and Michel Foucault (1926—1984). To them man's spiritual alienation from natural roots is in large part due to his focus on such anthropocentric constructs as religion, technology, society, culture and the like. Both argued, like Kierkegaard, that centrality of intellect in the historic Christian tradition is a major problem,[188] and a rebellion should be initiated and maintained. The mind must be overthrown, a sentiment promoted by the Green Dragon and nicely described by American deep ecologist Bill Devall, a disciple of Næss: "I think deep ecology is best expressed, not explained."[189]

It is clear the mainstream of modern environmentalism has, by the twenty-first century, imbibed rather freely of deep ecology. Mr. Mainstream Environment, Al Gore, is rather unoriginal as he channels the spirit of Heidegger, and Næss, page after page in *Earth in the Balance,* often mentioning how mankind is always trying, vainly, to think himself out of problems. Gore argues technology, or brainpower, is not the solution to ecological troubles, but part of the problem. Recourse must be had to "the part of our being that knows better…."[190] He is not talking, of course, about the intellect. One should mention that Al Gore's careful and logical business model shows that in his case avarice has overcome ideology.

Whereas deep ecology may have followers who are consciously atheist or pantheist, ecotheology is distinctly Christian in flavor. Roger Gottlieb, a distinguished professor of philosophy, explains in his anthology *This Sacred Earth: Religion, Nature, Environment*: "Ecotheology aims to reshape Christianity, by finding and stressing passages in classic texts that help us face the current crisis…. Contemporary ecotheology voices the sorrow of a broken-hearted earth and expresses our despair over the past and fear for the future. Simultaneously, theoreticians of religion and the environment question whether and in what ways they can connect religious energies to secular environmental philosophy and ecological activism."[191] Ecotheologians are thus more likely than deep ecologists to achieve rapprochement between environmentalism and Christianity. While deep ecologists are clambering over the gate, and perhaps smell funny, ecotheologians are already deep in the sheepfold and leading the bleating lambs out.

## Ecofeminism

In the history of the world the Earth Mother Goddess has featured prominently in various cultures under different pseudonyms, most prominently as Gaia, the primordial deity in the Greek pantheon. While some deep ecologists and ecotheologians will resent references to Gaia if intended as more than a didactic principle, for others the feminine connotations of "Gaia" are an essential part of their ecotheology. Feminist ecotheologians see environmentalism as the extension of the struggle against the androcentrism of traditional Judeo-Christian religions, which deep ecologists and others say is a major reason modern society faces a crisis of meaning.

It is no coincidence the rise of environmentalism as a significant political entity tracks the rising political clout of modern feminism. In ecofeminism one finds strong emphasis on healing, personal growth, authenticating experiences and self-realization. Here, as in the older models of the environmentalist movement, environmentalists reconceptualize God, as in Sallie McFague's *Models of God*, to be somewhat different from the God of the Biblical revelation, and not so different from Dewey's "idea." The Bible is no more than a useful reference point from which one may run a line to a "metaphorical theology." So-called monarchical models of God are passé, inauthentic, and must be overthrown. Models that promote the rights of Nature, the metaphorical "body of God," they therefore consider authentic. Ecotheologians revel in irrational positions. For example, McFague takes the position of pantheist mystics that "No human words can describe God."[192] McFague seems oblivious to the miracle she performed by using words to describe that which she asserted words could not describe.

Professing Christian environmentalists lightly toss about religious connotation words with little obvious concern for the accuracy of the ideas conveyed. "How vague your words are, Grandma!" says a modern little Green Riding Hood. "All the better to provoke emotional responses divorced from reason, my dear." If ecocentrism is the central organizing principle of the evolving faith these folks promote, then internal qualities such as self-affirmation and self-realization are the heart, argues Eccy De Jonge.[193]

While environmentalism can be ugly it is often just dangerously silly, such as among the chai latte sipping Hollywood elite who solve the problems of the world at yoga retreats and beach parties. Who better to describe the heart of self-affirmation and self-realization? Being Green is, like, the coolest fashion accessory ever.

This brings to mind Leonardo DiCaprio, who once lived a lust-filled life[194] but now, having found religion, occasionally flies commercial and has set up a Green foundation blessed by Al Gore. His eco-cred is up there with the best of superior scolds. Instead of a flashy sports car he crams himself into something smaller and slower, and feels good about it. The loss of sensual pleasure is more than compensated by supplies of spiritual pride. Or consider the sharp-shouldered imagery of ecologically sensitive celebrities Cameron Diaz and Gwyneth Paltrow chiding Joe Public in an advert to do as they say, presumably because they are cute. "Do you turn off the lights," they giggle, "when you leave your bedroom? … Do you turn your thermostat down to 65°F and wear a sweater? … Why don't you go buy a hybrid car?"[195] Though she wears only a flimsy peasant blouse and has very low body fat, Paltrow does not shiver but appears comfortable. Either she has ice in her blood or the studio thermostat was set above 65°F. Maybe she was demonstrating her willingness to be chilly for the sake of the planet by grinning and bearing it. Since that stretches credulity, it would appear some environmentalists fail to see the hypocrisy of preaching a "do as I say not as I do" theology.

## The End of Mind

When Al Gore, in his Nobel Speech, said, "Armed with such a faith, we might find it possible to resanctify the earth,"[196] he was speaking the jargon of ecotheology and deep ecology. He has become the public voice that has allowed more people to take seriously the strange morality of deep ecology and ecotheology, even if only a few radicals may subscribe to all of its utopian propositions and delusions. Although pagan spirituality is hostile to Biblical Christianity, it is precisely these ideas of deep ecology and ecotheology that have permeated society and the professing Church and facilitate a pagan-Christian environmental synthesis.

Oprah Winfrey is probably the world's most influential woman—tens of millions seek her guidance each week. She is also a professing Christian. Oprah channels her own brand of feminized spirituality through her show, calling it "televisual feminism."[197] Though she is no theologian, it is worthwhile to notice how her mysticism neatly meshes with that of contemporary deep ecologists and ecotheologians. Oprah recently began to promote vigorously the views of New Age guru Eckhart Tolle. Speaking of his spirituality she said, "God in the essence of all consciousness isn't something to believe, God is; and God is a feeling experience, not a believing experience…. If God for you is still about a belief, then it's not truly God."[198]

Summary: do not employ reason in the search to find something spiritually powerful and meaningful. Inspiration needs to come from somewhere deeper, a mysterious well of universal consciousness. But this is just what so many unwise Christian churches teach. Literal Biblical facts are not so important as vague metaphors and stories, which the naïve confuse with deep spirituality.

## Conclusion

Who then is to blame for the unsavory state of Western civilization, for the real sense of overwhelming crisis? There is enough blame to go around, but the churches bear primary responsibility, for they alone are the salt of the Earth. Those in the church who are not fundamentally dishonest have practiced a casual attitude towards the Bible, God's words, and so allowed all manner of mischief in the churches. When the church holds its mission—to preach the gospel of Jesus Christ—in low esteem, and prefers a false gospel, such as the gospel of Green, society will not enjoy the blessing of God. Nevertheless, His mercy and longsuffering are great, and He does not immediately plunge us into darkness.

For many professing Christians, the environmental gospel of Green is the carb-free synthetic secular substitute for their forgotten Christian faith. Though much of the environmentalist movement is rabidly antichristian, the religiously motivated environmentalists, who want an ecological reading of the Christian faith, are a growing constituency. However, the search for a Green Jesus, a vegan Earth Mother type, may be as successful as other such *faux* liberal movements. This is because the synthesis of environmentalism and Christianity is not saying anything unique among the secular mysticisms that surround us.

What impress the world are real mystics, like Eckhart Tolle, not Christian pretenders. The words of Presbyterian theologian J. Gresham Machen are fitting: "The truth is that the life-purpose of Jesus discovered by modern liberalism is not the life-purpose of the real Jesus, but merely represents those elements in the teaching of Jesus—isolated and misinterpreted—which happen to agree with the modern program."[199] Green Christianity is simply Christians conforming to the world spirit. It fits in nicely with the goals described by German Green politician Rudolf Bahro, for example. He says, "Our aim has to be the 'reconstruction of God'"[200] Under these circumstances, writes philosopher John Passmore (d. 2004): "I more than doubt Christian theology can thus reshape itself without ceasing to be distinctly Christian."[201]

The average person in the pews might not understand or have much interest in the philosophical principles that animate the Green Dragon, but this does not mean what they see and hear leaves them unmoved. Environmentalist propaganda constantly bombards them and tells them how selfish they are for not "caring" for all that environmentalists care about.

If foreswearing of intellect has become a chief feature of the Christian mainstream, then it will be a difficult task to convince professing Christians to pause and reconsider, in the light of the Scriptures, overtures from various guilt manipulators. Rather than "casting down arguments and every high thing that exalts itself against the knowledge of God, bringing every thought into captivity to the obedience of Christ" (2 Corinthians 10:5), they are having a mind job done on them at the bidding of the Green Dragon, by the hands of wicked and false shepherds. Mystical and emotional religion replaces revealed religion. The religiously motivated environmentalist, or the environmentally motivated Christian, has then placed himself in a dangerously unstable position and is ripe for easy exploitation.

Arguments that Christianity must be for environmentalism are only superficially Biblical; indeed, they are unbiblical. The Christian idea of stewardship gives man the responsibility to care for the creation and to use it wisely. Environmentalists reject this teaching in favor of the subtly different belief that the Earth's value is inherent to itself, that elephants, dolphins, trees, and viruses have no less intrinsic worth than human beings.

Insistence that all life is fundamentally one forms an important pillar of the movement. A logical result is a perverting of Christian respect for the creation into pagan reverence for nature. Some even believe the Earth is itself a living being. Religious fidelity to the cosmic biosphere is the central organizing principle binding diverse groups under the banner of the Green Dragon. The idea that man is the image of God they consider hopelessly arrogant and anthropocentric. At best man receives place as the nervous system of God[202]—that part of nature related to consciousness. Imagining that man can and should modify his environment to suit his wishes is bad. Thus progress as traditionally understood must also be rejected. We address these various environmentalist pillars in the following chapters.

Modern environmentalism is a reconstruction of a pagan world order, a rejection of Christian spirituality. Even so many churches are wittingly or unwittingly practicing the religion of environmentalism. Christianity is the narrow way, and God tolerates no idol competitors. Attempts by church leaders and Christian organizations to synthesize a Christian environmentalism succeed only by exorcising truth and, ultimately, expelling Christianity; all that remains is an empty chrysalis. Christ's church lives on, but not in the bare shell.

Ravening wolves in sheep's clothing have crept in and supplanted Christianity by a new religion that uses connotation language familiar to professing Christians. Not all Christians are fooled, and many are flowing out of the eco-congregations. This brings to mind the story of the gray wolf and the Inuit in the Canadian arctic. To kill the wolf the Inuit have devised a snare of sorts. They plunge a knife blade again and again into a bowl of blood. Because it is so cold the blood quickly freezes between dippings. Many frozen layers soon cover the blade completely, like so many strata of crusted makeup hiding the passing of time. Next the hunter buries the handle into the frozen ground, and leaves. Only the scent of death and the blood caked blade linger in the air.

Once the great gray wolf has caught the sickeningly sweet scent it draws closer to the blade, like the foolish young man to the painted and perfumed prostitute (Proverbs 9:13–18). One tentative lick is all it takes to inflame its blood lust. Soon the sharp edges of the blade become exposed and slice into the tongue. Yes, there is pain, even humiliation, but now there is also fresh, hot blood. So the wolf, green eyes blazing, goes on savaging the blade while its life wanes and drips into the snow. It slowly becomes weaker and weaker, staggers and chokes on its own blood, collapses, and finally dies.

Savage wolves have come to be among the church. Will it not be ironic if the Dragon they so adore ends up destroying them? Like their master their days too are numbered. No one can serve two masters (Matthew 6:24).

# CHAPTER THREE

# Created in the Image of God: Why Humans Are More Special to God than Other Creatures

*Oh, Adam was a gardener, and God who made him sees*
*That half a proper gardener's work is done upon his knees,*
*So when your work is finished, you can wash your hands and pray*
*For the Glory of the Garden that it may not pass away!*

Rudyard Kipling[203]

## The Beginning

Imagine a family lying on their backs in an open field beneath the vaulted sky. Under them is grass, fresh mown in the late afternoon. A dash of tender green moistness still hovers over the low-slung earth. Above their uplifted faces diamond-studded intensity perches delicately on the shoulders of a deepening black velvet night. Nearby, trees whisper on the wings of a quiet breeze. On one side, some distance away, rests a leaden heaviness of country mountains, sensed though not seen, silent watchers now veiled by inky darkness. Where did the stars come from? Why is the sky black as a wolf's mouth by night yet dazzling crystal by day? When did the mountains rise to kiss the sky? We humans tend to ask these kinds of questions. The course of civilizations depends, very much, upon the answers.

Our home, the Earth, is a majestic place. It formed about 4,600,000,000 years ago, an accident of the galactic flux. Like leaves racing rapidly on a star-spawned river, our atomic ancestors clumsily chanced together. Hopelessly, relentlessly drawn near, arching, twisting, spinning wildly over turbulent crest and trough, they collided in eager splashes of heat. Bound by gravitational glue in the drunken tango, gas and dust and planetesimals whirl in hypnotic harmony to the languid confluence of tides and currents and eddies that now rock them gently, not as impatiently as before.

In space, silence. Also, darkness and cold. It is so cold, about -450°F.[204] Below, there is growling, groaning, roaring, shuddering, as the cosmos gives birth to Earth. This is Gaia. Naked, and in the crisp coolness of the cosmic night, the new planet begins to shiver. Soon her ruddy form darkens, and hardens. A whimper from her cracking fragile skin shocks the evanescent sky. Sunlight caresses her and gently blows the afterbirth away from around Gaia, and as she bathes in golden rays she glories in her own unadorned elegance.

In good time she dresses herself with a thin mantle, the silvery atmosphere, to cover her shimmering beauty, and then ornaments of continents, and trees, and all creatures beautiful, wise, and wonderful. The little frogs croak at misty dawns, and tremble at their frog demons, and worship their frog gods. They live, and make love, and die, and evolve into greater frogs. And the greatest bug-eyed frog, so far, calls itself Man. And it preens and puffs proudly, and it paws Gaia irreverently, and imagines it is more excellent than she. But it is an ungrateful, slimy little frog perched precariously on Gaia's skin, confined beneath her mantle, easily consumed by her wrath.

Such is the history of man and Earth according to the environmentalist. Almost all religious and state schools, at all levels of education, teach a stiflingly similar history as gospel truth or, in more candid moments, as the best scientific guess of what happened before anyone was there to see. In academia it is interesting that the arts and softer sciences are most taken by the evolutionary gospel, ravished perhaps by the heady, cloying odors of the death spice. Scientists are slightly more skeptical. While they cannot return to the past, they may employ their inquisitive minds to imagine what it was like; observational evidence, logic, and imagination can go a long way. But in the end the best scientific theory of how man and the Earth came to be is only a tentative creation myth, stable as summer fog. Revelation alone can resolve a difficulty like this. So now, in this chapter, we turn to the Bible to see whether man is a frog or a prince, and what it means to be the image of God.

## The God Who Is Present

Human imagination breeds myths with a fecundity that shames rabbits. A ninth-century manuscript of the Jain religion, *The Mahapurana*, asserts: "Know that the world is uncreated, as time itself is, without beginning and end."[205] This pagan myth has much in common with the most modern twenty-first century scientific explanations that argue, to varying degrees, for a cyclic universe—in effect an everlasting recurrence of Groundhog Day, but set upon cosmic scales.[206] Modern people may flatter themselves they are not full of the ancient superstitions of their ancestors. Perhaps so, but they are not exceptional in their own mythmaking.

Creation myths, like the one mentioned above, litter all civilizations, for man is a creature uniquely curious about his origin. One could easily imagine a similar scene set almost 3,000 years ago, with a Babylonian family now lying in that dark field of green, inhaling sylvan silence and the stellar beauty of a deep night. To that Babylonian family, living in the time of great Nebuchadnezzar (about 630-562 BC), the best science of the day was clear on some answers, less on others. The supreme god Marduk used one half of the corpse of the rebellious goddess Tiamat to make the Earth from the primal matrix, the other half to make the sky in which he set starry stations for lesser gods. Marduk formed the mountains on the head of Tiamat, Mother Earth.

In her book, *Gaia and God: An Ecofeminist Theology of Earth Healing*, theologian Rosemary Ruether lists many such creation stories, including the Christian one. She writes: "Creation stories not only reflect current science, that is, the assumptions about the nature of the world, physical processes, and their relationships; but they are also blueprints for society … and … reflect the worldview of the culture and mandate that worldview to its ongoing heirs."[207] Though Ruether may be willing to say there is truth in the stories, for her they are never actually true. They never actually happened, but are always and only metaphors for a veiled past; they are true perhaps in the sense that Aesop's fables are true. But we are in a more favorable position than Ruether allows, because we have available a reliable written record of the true history of the Earth.

It is true that no man or woman lived to watch the moments of creation, but God was there, and in the Bible He tells us what He did, when, and even how He did it. God's Word makes us to be vicarious observers of what did happen, so we do not need to rely on imaginative myths inspired by pagan religions, or even modern scientific considerations. According to Ruether, as for Vanderbilt Divinity School theologian Sallie McFague,[208] the scientific myths are the strongest myths of our age. Maybe so, but one should not have to explain why strength does not require or imply that myths be true.

Many people believe that physicists can by unaided reason explain the cosmos; some physicists even believe it themselves. Yet without witness of the past the most excellent modern cosmological theories science can develop are just so many stimulating speculations. Philosopher Sir Karl Popper famously showed how the situation is even worse than many mythmakers suppose. He said, "First, although in science we do our best to find the truth, we are conscious of the fact that we can never be sure whether we have got it."[209] It is no shame that science is without legitimate authority of that which it cannot know, and it is even healthy to admit fundamental inabilities. Humility has its benefits, but God was there in the beginning, and He has made known what He did, so pretending otherwise is a false humility, and merely silly.

We live in an age that prefers symbolism over substance, sentimental or mystical appreciation of the spiritual rather than the immovable truths of the Bible. False theologians have led the way up the dunghill in the race to finesse the Scriptures into compatibility with Green political shibboleths; it is becoming something of a great tradition to rate going Green as a deeply meaningful spiritual orientation.

The game goes like this, explained in the words of ecotheologian Sallie McFague: "As I mentioned earlier, what can be said with certainty about Christian faith is very little; theology, at any rate, my kind of theology, is principally an elaboration of a few basic metaphors and models in an attempt to express the claim of Christianity in a powerful, comprehensive, and contemporary way."[210] Though she has suffered devastating critiques, for instance from Calvin Beisner,[211] McFague persists in relativist argumentation. She believes that no one has access to pure reality, yet publishes her beliefs with triumphant flair and without an inch of self-doubt. In his *Star of Redemption*, Jewish philosopher and theologian Franz Rosenzweig characterized such behavior as like a little child sticking her fingers in her ears while shouting "I can't hear you!" in the face of danger and death.[212]

## Deaf and Blind

God has spoken, but many professing Christians and Christian theologians like McFague cannot hear, or do not wish to hear, His message. We cannot say for certain, McFague says in effect, that Jesus Christ rose from the dead and will come again to judge the world in justice. She is all studied vagueness and doubt unless, of course, and oh so conveniently, she gets to decide what justice means. The current flavor of the month may be eco-justice, or feminist justice, or whatever political cause McFague thinks should receive her imprimatur. Why should anyone give serious attention to such irrational theologians whose God is a political cause? Flowing robes and emotionally touching words did not save the Pharisees from hell. Only the resurrected Christ saves sinners.

Important among a host of modern idolatries, the environmentalist thesis has enslaved minds, blinded eyes, and blocked ears. This is not unlike the situation recounted near the end of *The Chronicles of Narnia*. There a gaggle of dwarfs huddled in bright open air on clear, clean ground are so deluded as to think they are prisoners in a dank, dark stable. Though their creator had prepared for them a magnificent feast and stood in person but a breath from them, reason and evidence would not sway their fatal vanity. They ate the feast ungratefully. Gnawing delicacies with distaste, unable and unwilling to recognize quality, they munched in misery in a pathetic parody of precious half-remembered traditions. In rejecting their maker they did not achieve liberation, but created only a prison of darkness and misery. They became blind to goodness and truth.

Lucy, a queen in the story, expressed astonishment at their blindness. "But it isn't dark, you poor stupid dwarfs," said she. "Can't you see? Look up! Look round! Can't you see the sky and the trees and the flowers? Can't you see me?" But the dwarfs would not hear it, or of a personal God. They insulted the very *idea* in an ugly and pitiful reply, suggesting that such a God is a fraud: "How in the name of all Humbug can I see what ain't there? And how can I see you any more than you can see me in this pitch darkness?"[213]

Like the theologians, a few modern physicists,[214] such as Brian Swimme in *The Universe Is a Green Dragon*,[215] have portrayed matter and slightly more mystical qualities like energy in a spiritual ballet. Gary Zukav's wrinkled but timeless book *The Dancing Wu Li Masters: An Overview of the New Physics*[216] speaks about the cosmic ripples of organic energy and the quantum fields of green upon which moderns rest in trust as they gaze out into Milton's "wide womb of uncreated night."[217] Many in the humanities also love the vague religious coupling of science and spirituality, for it fits nicely in the relativist void where they have buried their intellects.

Environmentalists find great comfort in what they believe to be the increasing convergence of hard science and mysticism. In the book *Fashionable Nonsense*,[218] physicists Sokal and Bricmont ably lament and mock the pretentious abuse of science through science-words sprinkled on the nonsense spouted by postmodern intellectuals. By redefining words in their play world, religious environmentalists dream they are approaching God. They really approach an idol more in the image of Dewey's "unification of ideal values,"[219] Gore's "panreligious perspective,"[220] Tucker's "unified vision,"[221] or philosopher Erazim Kohák's "sacred presence."[222] Dwarfs they are, sitting in the palm of the Green Dragon, fondly stroking its upright fingers like bars of a cage through which they peer and jeer at those outside their prison.

They are prey of the Green Dragon, inebriated by at least the first fatal conceit of environmentalism. This lethal lie is to deny the existence of a personal God, or in the same breath to affirm that God is a mere force or energy, an idea or feeling, or meandering stream of consciousness. A great man like Al Gore feels able to both affirm and deny the existence of a personal God. He has affirmed that Jesus Christ is the one true God, but also that a new, seemingly pantheistic religion must redeem the world. To believe these at the same time is to deny the unity of truth—Jesus said that he alone is the way to God, and all other roads, howsoever gilded they appear, lead to hell.[223]

Logical lassitude in our moral and political leaders reveals either a lack of respect for reason and truth, or deep spiritual confusion—perhaps both. If the latter is accurate, one can only feel charitable sorrow for Gore's perilous spiritual torpor; if the former, one can only say that his honesty does not match his ability. Unfortunately, as previous chapters showed, a mélange of such chaotic and illogical ideas have great currency in modern culture. Even professing Christians have imbibed the foolishness, sometimes calling it deep or humble spirituality. But no sane person should admire the man who, after blinding himself, denies what he no longer is able to see.

What great sums of money the early Christian martyrs gave for a few pages of the Bible! It was not the paper, but the words of the Bible, that they valued. They loved its rebukes, its comforts, its warnings, all its excellencies. How precious in their sight were the words of God, because these words express His thoughts. Here is what the Bible has revealed: "In the beginning God created the heavens and the earth" (Genesis 1:1). This is the first trustworthy reference to time and space, from which we may learn the origin of the cosmos in the creative decree of God. Space and time had no prior existence until God spoke them into being. But God has Himself always existed, or perhaps a better way to state this is that God is eternal. All of space cannot contain Him, not even the heaven of heavens (1 Kings 8:27); He is the one true and living God.[224]

All other gods are fictitious; though it is possible to offer worship to the sun and other creatures, one wastes such worship on nothing but a glorious idol. No doubt statements like the preceding are offensive to the modern mind, but to the Christian they are both necessary and good. "The Bible and Christ in His office of prophet have," according to theologian Francis Schaeffer, founder of the *L'Abri Fellowship* in Switzerland, "given a propositional, verbalized communication to men that is true about God, true about history, and true about the cosmos."[225] To state or believe otherwise is to treat God's revelation as much less than the Word of God, and no more than the passing passé myths of a few ancient nomads. Is this what the Bible is?

To the unbelieving modern mind the rare, uncomfortable interaction with genuine Christian ardor, constancy, and inflexible perseverance leaves the distinct impression of an encounter with something at once primitive, extravagant, and absolutely outrageous. But in point of fact, modern pagans on the left and right are no less dogmatic, and may be much more so; to them Christianity is at best an agreeable illusion to humor when expedient. Consider the late great Carl Sagan, a much-respected and honored space physicist, who wondered[226] how those who ride airplanes can believe in God or demons.

In his influential book *Cosmos*, Sagan wrote, "The Cosmos is all that is or ever was or ever will be."[227] Obstinate and fond assertions like these are sadly common among even outstanding scientists, as also among Hindus, Baha'i, Wiccans, indeed a veritable buffet of pagans. None of this should be understood to suggest that scientists are necessarily pagan, or that pagans are scientific, just that all humans are inescapably religious. It does suggest that embarrassment or timidity about Bible truth, truth ratified by Jesus Christ's resurrection, is unbecoming of Christians. Not only does God exist, but He created the heavens and the Earth, in the beginning, and out of nothing. He was *personally* present, He was active, and He is still there and still active.

## Dust of the Earth

Unlike God, the creation is finite, for it has a beginning. Thus the created order is not a part of the God who made it. Rocks, humans, angels, and stars depend upon His pleasure for existence, and came into being by His free choice.[228] The Holy Spirit, who directed the tongue of the prophet Isaiah, clearly shows how God rules everything: "Thus says the LORD, your Redeemer, and He who formed you from the womb: 'I am the Lord, who makes all things, who stretches out the heavens all alone, who spreads abroad the earth by Myself'" (Isaiah 44:24). Everything is subject to God, argues Isaiah, and the demonstrations of God's power show that His words are trustworthy.

In Job 22, it is written, "Can a man be profitable to God, though he who is wise may be profitable to himself? Is it any pleasure to the Almighty that you are righteous? Or is it gain to Him that you make your ways blameless?" The answer to these rhetorical questions is, of course, No. In short, God needs nothing from us, or from the cosmos. His will is not subject to coercion. We, on the other hand, are never free from the providence of God; all our actions are contingent.

The first few chapters in the Book of Genesis are an infinite mine to plumb for riches. All the world has no wisdom that is greater. When the Bible teaches that God spoke, as in Genesis 1, "Then God said …" (v. 3)—it reveals that God spoke the entire creation into existence by His Word. Not only does He communicate, but He uses words, not just symbols or pantomime. Of course, the Scriptures contain some things that are difficult to understand, some that are mysterious or perplexing. Deuteronomy 29:29 says, "The secret things belong to the LORD our God, but those things which are revealed belong to us and to our children forever, that we may do all the words of this law." Considering its Author, it is only natural that the Bible contains some deep and unfathomable mysteries. But this absolves no one from negligence of the truth and obedience to those things God reveals. The absence of exhaustive knowledge does not imply the absence of knowledge.

Mary, the mother of Jesus, faced the mighty angel Gabriel, who brought to her word from the Lord God that she would conceive a Son, though she was a virgin, and that this Son would reign as king forever. While our encounters with the messengers of God are not so dramatic, we are, each one, still faced with Mary's choice: to believe, or to reject the words of God. It is true that Mary was young, humble of means, and apparently poorly educated. Yet this was not a disadvantage in her most important decision. Her cousin Elizabeth, filled with the Holy Ghost, remarked, "Blessed is she that believed" (Luke 1:45). Mary heard God's word, spoken to her by an angelic tongue, and though astounded she believed. Unlike Mary, many professing Christians choose in place of God's Word experience, subtlety, tradition, science falsely so called, church or political authority. They despise God's Word. It is against goodness and reason that any of God's creatures should despise His Word.

Physicists study God's world and, of all people, know (or ought to know) how shaky is scientific knowledge. In one age Newton was almost godlike in his prescience, but he is merely a useful genius the next, his theories superseded. It cannot be helped; it is our nature to be often revising our opinions. But God does not change. Although God is infinite and we are finite, and our knowledge must therefore also be finite, the knowledge we have from God in the Bible is true knowledge, for the Word of God is, as evangelist George Whitefield (1714—1770) put it, "the great Fountain of truth."[229]

Seeing as God is so utterly different from humans, one might perhaps excuse a momentary faintness of mind on catching a second sinuous lie bred by the Green Dragon: there is nothing special about humans. But astonishment, like lightning, lasts only a moment, for the Bible speaks and teaches the clear antithesis (Genesis 1:27): "So God created man in His own image; in the image of God He created him; male and female He created them." After He made all other creatures, only then was it time to reveal the image or likeness of God, the glorious crescendo to crown God's cosmic masterpiece.

God's choice to make mankind His supreme creation, the last thing He made, at once marks humans as extraordinary creatures. When God said (Genesis 1:26), "Let Us make man in Our image, according to Our likeness" there was clear internal communion of agreement in the Godhead. Yet it is not as if this consultation is the result of a sudden apprehension or need. Rather, God would have humans to know the dignity of their being by the special care He takes in creating man. Well He knows our feeble frame and so provides a word to help our ignorance.

Sixteenth-century Reformer John Calvin noted, "For God is not now first beginning to consider what form he will give to man, and with what endowments it would be fitting to adorn him, nor is he pausing as over a work of difficulty: but, just as we have before observed, that the creation of the world was distributed over six days, for our sake, to the end that our minds might the more easily be retained in the meditation of God's works: so now, for the purpose of commending to our attention the dignity of our nature, he, in taking counsel concerning the creation of man, testifies that he is about to undertake something great and wonderful."[230]

Amid the creative glory revealed in God's Word, the claims of the Green Dragon melt away into silliness, like an ice cube on the sun. Not that this stops the Green Dragon from contradicting the Word of God. Peter Singer, the influential Princeton University animal-rights philosopher, spells out what the gullible are slowly beginning to realize through painful experience: "When we reject belief in God we must give up the idea that life on this planet has some preordained meaning. Life *as a whole* has no meaning. Life began, as the best available theories tell us, in a chance combination of gasses; it then evolved through random mutation and natural selection. All this just happened; it did not happen to any overall purpose. Now that it has resulted in the existence of beings that prefer some states of affairs to others, however, it may be possible for particular lives to be meaningful. In this sense some atheists can find meaning in life."[231]

Meister Singer in effect tells us to raise our hands to cover the sun, and—voila!—it is no longer there. And he is, in a sense, right. But he also wants to pretend that the act of covering the light with his hand gives meaning to life. It is hard to take him seriously, since it is not clear whether he takes himself seriously, or is merely mad. If Peter Singer, or any other man, is the standard by which to judge all else—and Peter Singer knows we are smaller than a speck of dust when viewed from the Space Shuttle, or from a sun orbiting satellite—then what is man?

And yet for all the big, bad wolf's huffing and puffing it is not possible to conceal man's true greatness and nobility, for he reflects, even if only in some veiled way, the glorious image of the great God who made all the planets and stars and galaxies in the cosmos. This is heady fare.

Lest their preeminence in the celestial symphony of creation become a temptation to pride, the Bible quickly moves to show that man is a composite creature. In Genesis 2:7 Moses elucidates further details of Adam's origin: "And the LORD God formed man of the dust of the ground, and breathed into his nostrils the breath of life; and man became a living being." God is a spirit and the image of God "consisted partly in the spirituality of the soul of man."[232] There is a special glory in being the image of God, but also a certain measure of lowliness, in that inanimate dust of the Earth forms his flesh. A recurring theme in the teachings of the Green Dragon is slightly to stretch, or crimp, the real state of things. We are, as Singer so pointedly teaches, made of dirt, but we are oh so much more, as the Bible teaches.

## Reason

God formed the woman not long after Adam but in a different way. After anesthetizing the man, God opened his chest, extracted a rib, and used this as the matrix of cells from which he formed for him a suitable helper. In the original Hebrew the construction work is still more clear, for it says that "the rib he built up into a woman."[233] In some sense then, as stem cell research currently proves, one may arguably consider the woman a clone of Adam; she reflected Adam's likeness and shared his genetic identity.

Later God brought her to Adam and gave her to him as wife, acting as parent of both bride and groom. On seeing her for the first time, Adam exclaimed in delight (Genesis 2:23), "This is now bone of my bones and flesh of my flesh; she shall be called Woman, because she was taken out of Man." She was his mature counterpart, the opposite that completed his existence; she was his flesh. Scripture teaches that even though we have in common with animals a physical body, still "[a]ll flesh is not the same flesh, but there is one kind of flesh of men, another flesh of animals, another of fish, and another of birds" (1 Corinthians 15:39). Though the substance was the same, there was a spiritual and material quality that set man apart from other creatures, a difference recognized by Adam.

Adam's exclamation was more scientifically correct than he could have known and should be, at least in principle, respectable to even the most atheistic person who yet has faith in modern scientific ideas. Now that Adam had his Eve, a creature closer to his nature as made also in God's image, he was able to enjoy conversation and other social intercourse with one of his own kind. He had previously listed the astonishing diversity of the other creatures, but it must have become clear that though each one had its own mate, yet not one animal could do for Adam what Eve could do. Adam recognized in his exclamation that she was the helper that suited his nature flawlessly, to whom he would cleave in deep, personal, and mysterious communion. He not only knew God, plus his relation and duty to God, but he knew himself. He was not an infant mind trapped in an adult body but a mature person, body and mind.

In all these passages in Genesis, verse after verse makes it clear that Adam, like God, partook of speech and reason. Like God Adam was capable also of deliberation. Indeed, this was already plain before Eve's construction. God had previously given Adam various commands, such as to dress and keep one special place on Earth—a garden called Eden (Genesis 2:8,15). He was to refrain from eating the fruit of one tree in the garden (2:17), and like a king to survey and name all creatures paraded before him (2:19). In the first systematic book on science published in America, *The Christian Philosopher*, Puritan minister Cotton Mather (1663—1728) justly calls man the "lord of this lower world"[234] because of his preeminence. In exercising this lordship Adam reflected the image of his creator, who is the King of kings and Lord of lords. Man, and the woman made after his image, and therefore the image of God, is not a dumb-as-mud automaton but a miraculous creature filled with sublime reason.

In 2 Peter 2:12 Scripture signals that man has reason far above the dumb animals. It says that when people hate God and are slaves to their own lusts they behave like animals: "But these, like natural brute beasts made to be caught and destroyed, speak evil of the things they do not understand, and will utterly perish in their own corruption." Scripture marks the animals as void of manlike reason, following only the gross instinct of their nature. This is not to say that after their own way animals do not plot or scheme. They are, however, governed not by religion, but by their own fleshly desires.

Presbyterian theologian and Bible commentator Albert Barnes (1798—1870) wrote, "Man has reason, and should allow it to control his passions; the brutes have no rational nature, and it is to be expected that they will act out their propensities without restraint. Man, as an animal, has many passions and appetites resembling those of the brute creation, but he is also endowed with a higher nature, which is designed to regulate and control his inferior propensities, and to keep them in subordination to the requirements of law. If a man sinks himself to the level of brutes, he must expect to be treated like brutes; and as wild and savage animals—lions, and panthers, and wolves, and bears—are regarded as dangerous, and as 'made to be taken and destroyed,' so the same destiny must come upon men who make themselves like them."[235]

Remarkable among animals is their lack of history. They do not care who their ancestors were, or where they lived. Animals invariably live for the moment in a government by instincts and hungers. It often appears that for animals the only meaning or significance to events is whether they lead to a meal, or satisfy the other instincts. Man is not an animal, since he lives in time, and extrapolates connections between the past and the future, all guided by the providence of God.

The Green Dragon, like Nietzsche,[236] would have us deconstruct the human being and his civilization into pure animality. If a nod must be given to spirituality, then either a Spinozan or pagan pantheism is preferred. Environmentalists often stress the continuity between animal and political life, explaining behavior by animalistic evolutionary imperatives. However, a mind freed from the stifling bonds of evolutionary dogma can see clearly that, except in extreme poverty, pain, and fear, where mere existence is a massive challenge, human beings gravitate towards higher-order motivations. And even in extreme cases the astounding nobility of man regularly rises to prominence. Mortality is too frail to restrain the mind of man.

Perhaps the greatest capacity of man, most distinguishing him from the beasts, is the spiritual capacity, "because what may be known of God is manifest in them, for God has shown it to them" (Romans 1:19). In this we learn that man has the capacity to know there is a God, as many ancient philosophers have concluded. As man watches the world his spiritual senses tingle. Peter Singer speaks of his own experience of the wilderness beauty in explicitly religious terms. He writes, "I do not think I am alone in this; for many people, wilderness is the source of the greatest feelings of aesthetic appreciation, rising to an almost spiritual intensity."[237] The spiritual appetite is an indelible imprint on the lives of men. Scripture tells us, the "law [is] written in their hearts" (Romans 2:15). In other words God has gifted us with spiritual powers and faculties that enable us to see right from wrong and, depending on our actions, approve or accuse our conscience. What Peter Singer expresses is his discovery of the hidden spiritual feature of his heart. He thought it lifeless and buried, but it will not be denied; deep calls to deep, for God has "put eternity in their hearts" (Ecclesiastes 3:11). If we were not so naturally sinful, our lives would be ones of continual wonder and praise at our observations of the glory of God.

"What a piece of work is a man!" cried the almost madman Hamlet, "how noble in reason! how infinite in faculties! in form and moving how express and admirable! in action how like an angel! in apprehension how like a god! the beauty of the world! the paragon of animals! And yet to me what is this quintessence of dust?"[238] It is impossible for animals to produce a Shakespeare because for the animal everything changes continuously; words and tones are just compressional waveforms sensed by their organs and associated with certain events. There is no cultural memory, no past, no hope for the future. Their knowledge of God, if one could call it knowledge, is on a purely physical level. For man, who knows in his spirit that God does not change (even if he will not admit it), comprehension of deep mysteries is possible.

Jacques Derrida discusses a few features that distinguish man from animals, including "speech, reason, experience of death, mourning, culture, institution, technics, clothing, lie, pretense of pretense, covering of tracks, gift, laughter, tears, respect, and so on."[239] These distinctions do not deny that some animals, elephants for instance, display evidence of complex thought in their seeming grief for dead of their kind. Elephants appear to have a deep need to remember and mourn dead clan members, often stopping or pausing at the site where a comrade died. Dogs, too, are man's best friend for a reason. But the presence of complex capacities, even though limited, is not a good reason to be blind to the nearly infinite gulf separating man and beast in the high-order abilities.

Mankind alone among all creatures in the known universe holds awe-inspiring reasoning gifts. It is true that among men multitudes behave just like the beasts. They do not rise above their animal passions to contemplate intelligently the higher order of religion. But here we are speaking of Adam. Unlike us, much of Adam's knowledge was impressed upon him at his creation. Much of what we know we acquire through slow experience, but from the beginning of his existence Adam spoke and reasoned with maturity; much of his knowledge was unlearned. It was not animal instinct that he had, but a primary knowledge that God passed on as a special gift.

Adam did not just name the other creatures, including the woman, but the names he chose were truly descriptive of the nature of the other creatures. When he gave to the woman the name *Eve* (Genesis 3:20), which in the Hebrew means life-giver, it was clearly an apt choice, since she did just that through childbearing. Our first parents were able to understand cause and effect, the true nature of the world, and to express their ideas in language.

Two parts, equally clear, make up all mankind's nature. One part, the soul, is spiritual, and the other part, the body, is purely material. Both are indispensable to complete the whole man. Magnificent and indispensable though the body is, it cannot compare in sublime production to the human mind. The polymath Blaise Pascal (1623—1662) said, "Thought constitutes man's greatness."[240] Here, in the mind of Adam, the magnificent gift of reason reveals a chief part of the image of God. Man's reasonable nature is a facet of his person, much used in matters of religion. However, as was soon to become obvious, Adam's reason was finite and subject to subversion, so this teaches us how important it is to seek submission to God's revealed Word (Colossians 3:10).

## Righteousness

Righteousness is another prominent attribute of God. Scripture refers to God's righteousness a multitude of times, such as in Psalm 145:17, which is typical: "The Lord is righteous in all His ways, and holy in all His works." To say that God is righteous means that He always does right and is impartial and unprejudiced in all His judgments. The word is closely associated with justice, though little to do with the justice of the Green Dragon, whose tender mercies are a refined byword for cruelty. To be right is to conform to God's law. God is therefore perfectly righteous; in Him there is no shadow of turning (James 1:17).

The Bible speaks of humans as righteous too, but also as desperately wicked. For example, when God determined to destroy the degenerate Sodomites "because their sin is very grave" (Genesis 18:20), He chose to spare Lot, a righteous man. The righteousness of Lot, and of all true followers of God, is lastly not personal but derived from faith in the Savior (Romans 3:20–27). Before that fiery cataclysm Lot's cousin, Abraham, begged God to spare the city for the sake of but ten righteous in it. Abraham appealed to God's own righteousness: "Shall not the Judge of all the earth do right?" (18:25), and God acceded. In the same book God reveals His good and righteous nature again and again. He desires that humans be righteous, saying in Jeremiah 9:24, "I am the Lord, exercising lovingkindness, judgment, and righteousness in the earth. For in these I delight."

That God is righteous points to the evenhanded fairness with which He judges all individuals. Not to pick too much on the Sodomite scab, but their desperate plight shows that at a minimum righteousness requires moral and personal separation from wickedness. It also illustrates that all God does is right whether or not it is right in the eyes of man. The Sodomites were modern in their contempt for God's standards of goodness, for instance in their bustling efforts to convert Lot's front step into a gay bathhouse. But the wickedness of Sodom grieved righteous Lot, says Scripture. When a person submits to God's Word, he is obedient to God. Such obedience to God is but another name for righteousness.

I do not need to argue that mankind is now unrighteous, as that is obvious to all but the most deluded. The Bible teaches that God wrote His law on each heart, just as it was on the hearts of Adam and Eve, and that we choose to sin (Romans 3:10). This will to sin, to revolt against the law of God, is the extreme opposite of righteousness. But unlike us, who naturally cannot but sin, Adam had a natural will to do goodness. We know this because the Bible teaches that Adam was righteous, for instance in Ecclesiastes 7:29: "Truly, this only I have found: That God made man upright, but they have sought out many schemes." This verse covers the gamut of humanity, first to last. At the first Adam was without fault and perfect in righteousness, but after he turned aside sin became an enduring thorn in his side, as it is for us. Even on good days it is a constant struggle, even a bloody spiritual war to the death, to think, say, and do what is right. Adam's original righteousness was a pure native reflection of the righteousness of God, a radiant part of the image of God in man. How excellent it must have been to have no concept or knowledge of sin, to be free of struggles with evil, even if only for a brief time!

Adam had a perfect righteousness and, almost beyond belief, we too may have it, and more. Though we do not come into this world with the will to do right, like Adam had, yet we may receive as a gift more than that original native righteousness of Adam, through the man Jesus Christ. Because He is fully God, the righteousness of Christ is infinite, unlike the original righteousness of Adam, which though perfect could never be infinite. This is why Christians frequently call Jesus the second Adam, and a greater than Adam (1 Corinthians 15:47). Through Him it is possible to recover this element of the original image of God, as Ephesians 4:24 says: "the new man … was created according to God, in true righteousness and holiness." To become a follower of Christ is therefore to receive anew the image of God in righteousness.

Righteousness is so important an attribute of God that He binds His name to it. For example, in Jeremiah 23:6 God's people hear, "this is His name by which He will be called: THE LORD OUR RIGHTEOUSNESS." Jeremiah lived in a time when the people of God were enormously degenerate, and committed to false worship. Since worship is arguably the highest expression of man's relationship to God, the unfaithful behavior of Israel he likened to the gyrations of a whore. Nevertheless, his saying the name of the God of Israel is OUR RIGHTEOUSNESS shows that the undeserving people of God have righteousness in common with God. This is only possible because Jesus, as God, possesses perfect righteousness and gives it as a gift to those He loves. To the Christian the Lord Jesus, the second Adam, "who is our righteousness, is God manifested in the flesh."[241]

Scripture teaches that righteousness is as pure a product of true love as exists. "For this is the love of God," says the apostle John, "that we keep His commandments. And His commandments are not burdensome" (1 John 5:3). In the strange parallel universe occupied by the radical environmentalists we have been quoting righteousness has little or nothing to do with God, except as a tool to manipulate or force others to do things their way.

When ecotheologians couch their sermons in spiritual terms such as "righteousness" or "justice" they practice deceit, for they use the words with precious little reference to the Bible's definition. Liberal theologians have little problem appreciating this argument, for they, as Al Gore has written in a different context, recognize how politicians are "invoking the language and symbols of religion to bypass reason."[242] Nietzsche, the philosopher destroyer, was an equal opportunity offender, attacking hypocrites wherever he saw them. Though perhaps referring to the socialist rabble he so despised, he said something equally applicable to environmentalist politicians and theologians: "Distrust all those who talk much of their justice! Verily, in their souls not only honey is lacking. And when they call themselves 'the good and just,' forget not, that for them to be Pharisees, nothing is lacking but—power!"[243]

## Holiness

Words cannot describe how different God was from Adam, or from us. Though Adam was God's image, yet God is and was utterly unlike man in the immensity, and especially the incomprehensibility, of many of His attributes. Our knowledge as finite creatures is feeble, and incapable of encompassing God's infinities. But rather than humbly submitting to the teachings of God found in the Bible, and much in the same vein as twentieth-century liberal theologians, some ecotheologians use native human inability as a foil to smuggle into the church the mysticisms of the Green Dragon. This is as desperate as the person who thinks he needs to shoot up on drugs because life is hard to understand.

Like sentient slugs who wish to wrap their slimy trail around the world, ecotheologians have slouched in exhaustion to the top of the dunghill of liberal mysticism, only to focus anguished, perplexed eyes on the vertical wall of Mount Everest rising before them. God is utterly pure and separate in His holiness, a mountain no one can climb. Faced with the full authority of the Scriptures, they find they have nothing worthwhile to stand on; yet they are unwilling to abandon their dunghill. As when a car slams into a brick wall, they need to recognize their worldview is bankrupt, and stop and turn around.

Otherwise when the infinite wall of God's truth presses on the limits of comprehension irrational despair might seem like a logical choice.

Imagination does play an important role in Christian faith, allowing true believers to internalize the verbal propositions of God's Word and to realize that which is pure spirit. But imagination must always be subject to the Word. For instance, Scripture is replete with passages in which God, who is a pure spirit and whom heaven cannot contain (1 Kings 8:27), is said to have body parts. God helps our feeble imaginations by providing symbols and illustrations of the spiritual realities so important to our existence. Imagination unhinged from the soundings of God's Word will founder as certainly as a ship guided by a blind hand will run aground.

Human reason, because of its exalted position, must also be subordinate to the Word of Christ, or risk still more terrible ruin. As G. K. Chesterton opined, "Poets do not go mad; but chess players do. Mathematicians go mad, and cashiers; but creative artists very seldom. I am not attacking logic: I only say that this danger does lie in logic, not in imagination."[244]

God is, of course, logical, though we in our fallibility often mistakenly think not. He commands His servants to reason, to think, to consider, and weigh carefully. This is perfectly rational and reasonable, and contrary to mindless mysticism. And, as is the case for imagination, divine logic prostrates autonomous human reason before God, and before the eternal and holy mountain of His Scriptures. A proud man may become upset by expressions of the Biblical Christian position, for it informs him not only that he is barred from certain secrets, but also that there are many others he is incapable of grasping. To him it is a psychic barb to hear that "the foolishness of God is wiser than men, and the weakness of God is stronger than men" (1 Corinthians 1:25). He would rather face destruction, like Pharaoh of old, than bend the knee and bow his reason to God. Such recklessness before a mighty king can only be madness. The Bible calls it foolish.

There is in the embrace of a Green God something of the lunatic logic of a Ludwig Boltzmann or a George Cantor, respectively a spectacular nineteenth-century physicist and a mathematician, both driven mad through their studies. Opponents apparently drove Boltzmann to suicide by their insistence that atoms did not really exist, only energy. And Cantor—well, he studied infinity without God and could not handle it.

Engaging with reality may be painful, but blind faith is an irrational alternative. God is a fearsome God, but He is not capricious. He may be terrible in His awesome holiness, but He is also very good. To finite minds the infinite mysteries of almighty God are no doubt perplexing, but to embrace God is to embrace life. Embracing the Green Dragon is, in a real sense, intellectual and spiritual suicide. Psalm 2 gives sage advice to kings, just as applicable to the man on the street: "Serve the LORD with fear, and rejoice with trembling. Kiss the Son, lest He be angry, and you perish in the way, when His wrath is kindled but a little. Blessed are all those who put their trust in Him."

God is also internally self-consistent. I mean that He is—and I tread lightly here—in some sense self-consistent in the manner of the self-consistency met in physics, for example in electrodynamics. In the physics sense, self-consistency means there is no internal tension within the particular body of knowledge; every statement, every prediction, is a rigorous logical construct that is in perfect harmony with all the known principles of the theory. If the metaphor is useful, we may say that God, in this sense too, does not contradict Himself. However, an important difference is that unlike physical theories, which change not because logic changes but because new discoveries sometimes falsify old ideas, providing new premises from which new conclusions must derive, God does not change because God is truth; His premises never change. The process of logic does not change, yet logical conclusions change continuously as do the scientific theories that provide their premises.

God never changes; He is altogether *exalted* and *different* from any of His creatures. The seraphim exalt God in crying, "Holy, holy, holy is the LORD of hosts" (Isaiah 6:3). God singles this attribute out from His nature as one upon which He swears to perform whatever He has spoken (Psalm 89:35). He is holy in the past, holy in the present, and holy in the future. When faced with the incomprehensible holiness of God men have before them two serious choices. The first is to run from the light, but this is madness since to run that way will lead to a certain plunge into the fetid darkness of some or other idol. The second is to fall on one's face and acknowledge that God is great, and man is dust and ashes.

The Bible carefully develops the meaning of holiness, especially in the system of worship revealed to Moses. When Moses first saw the miraculous bush that burned without being consumed (Exodus 3:5), God commanded that he put off his shoes. Through the sensual experience of this outward action God excited Moses to an appreciation that he was in the presence of God, before whom he must abase himself. He was reminded of the holiness of the ground, which he was to walk upon with great care. Ground that is holy is different from ordinary ground. On holy ground one must tread lightly.

Not that the stuff of the ground was intrinsically holy. Moses may have walked on that same nearly bald patch of earth before, and not given any special reverence to the place. We might walk on the spot today and not know it or receive any special blessing. No, it was the special presence of God that made it out of the ordinary, holy.

Holiness is the setting apart of a thing to use it to the glory of God. But God decides what is, and what is not, holy. When man decides, it always leads to sin. Even a shallow examination of the Bible reveals that holiness is the opposite of sin. For Adam, in his first sinless state, the air that he breathed, the ground he trod, the stars he gazed upon, the breeze on his skin, the heat, the cool—everything was holy. In the words of nineteenth-century theologian James Thornwell, "Nature was a vast mirror, reflecting the Divine beauty, and as he saw he loved and adored."[245]

Adam too was holy; holy because he delighted in God. He was not afraid, far from it. He appeared at ease in the presence of the thrice-holy God. God set him apart. Of all the creatures God walked and communed with him only. This is what holiness is, being set apart to the Lord. Adam was all this, as he for a brief time reflected the image of God in perfection.

## Broken Clay

Adam, who is then the image of God and can see, comprehend, and reason, knew truly the source of the woman. He named the creatures and so described their natures. This shows that he was able to grasp the true nature of what he saw and understood, and that he could express that nature. God set him apart also from the other creatures in the mode of his creation, his role towards other creatures, and the means by which God provided him with a helper. So the man not only could come to true knowledge, but was also holy to the Lord. Finally, God declared him to be very good, thus righteous.

Adam and Eve were capable of volition, having the liberty to choose obedience to God's command and life, or to choose disobedience and death. They chose death. God made Adam and Eve, created in His image and so mirroring His glory in their perfect and sinless knowledge, righteousness, and holiness. But it did not last. When Adam chose to sin, not only he was cursed, but also his wife, his posterity, and all the creation. Adam, Eve, and their posterity were stripped of the holiness and righteousness in which they were formerly clothed.

It is needless to point out that modern man is unholy as he is unrighteous. Walk through the downtown area of any major city and one can find filth and trouble without much effort. People are ever ready to take advantage of even the weakest members of society. Even the nicest person struggles in his heart to do what is right and holy. This is an appalling position for us to be in, for without holiness "no one will see the Lord" (Hebrews 12:14), and "'There is no peace,' says the LORD, 'for the wicked'" (Isaiah 48:22).

We may be smart and still bear some small ability to reason, but humans are most certainly no longer good, not since Adam. God's Word does not shy away from describing the terrible unrighteousness found among even Christian heroes. In just one family, that of the patriarch Jacob, from whom the nation of Israel took its name, it seems that all the multitudes of vicious sins congregated. Their sins, which like venom bubble up to stain the pages of Genesis, are far more numerous than one can list. In only one generation idolatry, lying, theft, rape, dishonoring of parents, murder, kidnapping, adultery, and prostitution are just a few of the hellish barbs that spit cruel shame and disgrace on the name of Israel. We are not so different from them.

It is good for our race that Jesus said, "I have not come to call the righteous, but sinners, to repentance" (Luke 5:32), because if only the righteous could go to heaven then no one could go. The painful reality of our broken world does not lend itself to serious motions of confidence in the future perfectibility of human nature, however much utopian humanists—Green or red—may hum its dreamy tune. Atheistic secular humanism is the most destructive religion in all of history, having made the twentieth century killing fields the bloodiest of all time. It is scarcely possible to imagine a just and wise person being willing to die to save supermen like Hitler or Stalin or Mao or Mugabe. Who would be willing to die for a person who wanted to harm us, destroy our reputation, or corrupt and destroy our family?

From Adam to Jesus, up till our day, these two only were obedient to perfection and thus truly righteous, and Adam did not remain so. He fell fast from his perfect righteousness. Jesus remains as the lone tree standing in magnificent perfection on the mountaintop. Jesus was conceived and born in purity, without any remnant of Adam's smut, and Jesus lived and died perfectly. If not for Him, the name *Israel* would be yet another byword for the filth of the world, and the ugliness of sin. He redeemed the name, purified it from its pollution, and invited all nations to join in the mighty salvation, as a graft into the tree of life. We too may thus live with Him on the heights of grace and glory.

## Grave to Grace

When we are over familiar with our lover it is easy to neglect details, to know her face so well from a distance that we know nothing of the details. We think we know her so well. But would we ever recognize her ear in a crowd, or the difference in skin tone or texture on a part of her face? Would we know him by a touch in the dark? Loving him again, from a closer distance, can be a startling reawakening. Magnification and concentration of attention teaches us of places with new lines that we never before imagined. John 3:16 is like this. It teaches a most remarkable truth, one often overlooked because the verse has been so abused as to become commonplace.

It says, "For God so loved the world that He gave His only begotten Son, that whoever believes in Him should not perish but have everlasting life." Notice that God first loved our race and then, based on this love, gave His only begotten Son to bear the punishment for sins. He did not send Christ, His only begotten son, and only then love sinners. No, the first cause of salvation rests entirely on his eternal electing love. As it says in another place, "He chose us in Him before the foundation of the world, that we should be holy and without blame before Him in love" (Ephesians 1:4).

It was therefore necessary that we humans bear the image of God. It is not for angels or any other creature, but for man alone, to be so marked with heavenly honor. If God is the God He says He is—and this we believe is true—then His love is entire and complete. It will not suffer addition or subtraction, no variation in shading or depth. There comes no change in His perfections through His setting eternal love upon sinners. His perfections only become plainer to us; as our understanding of God grows, He grows more and more glorious in our eyes.

Yet this is possible only when we come into glorious communion and union with Him. It is not that humans become like God, which is the eternal lust of Satan, but that we are the eternal reflection of God. An image in a mirror adds nothing to the first cause, but allows observers to become aware of the present glory of the original being. So too we humans are the image of God and add nothing to His eternally infinite glory. We are part of His glory as Adam was in particular (1 Corinthians 11:7), but only because we are mysteriously linked to God as His express image.

Further, Scripture teaches that "'The first man Adam became a living being.' The last Adam became a life-giving spirit," and "as we have borne the image of the man of dust, we shall also bear the image of the heavenly Man" (1 Corinthians 15:45, 49). This means that man has not only a natural, or physical, body but also a spiritual body. Adam unites every human in the flesh, and every human receives in that flesh the suffering and death bequeathed by that first Adam. Our material bodies, though buried and rotting in the earth just like the carcass of a cow or a rat, will, unlike the beasts, rest in hope.

As verse 49 teaches, however, the story does not end there in the grave. Those called to embrace Christ are born again after the heavenly image, namely Jesus, the last Adam. Christians become quickened, or made alive again, by receiving a new nature. As death approaches, the believer receives comfort in the sure knowledge that not even death has power to sever his relationship with God. Indeed, to be absent from the body is to be present with the Lord (2 Corinthians 5:8). And when the body is dead, it will dissolve like a seed, and then burst into glorious new life at the resurrection. Christians will receive a new body when they pass through the gate to glory and reunite with their now perfect spirits.

Verse 49 also means Christ was made flesh, with the same visible image we have, but perfect. Though our nature is so far below His divine nature, even infinitely below, He chose to live among us so we might again be able to glorify and enjoy Him, and to live with Him forever. He adopted Adam's image, our nature but for sin, and denied Himself no humiliation. No pain and suffering was too great for Him, that no pleasure should be withheld from us. Marvelous wonder of redeeming love! "For scarcely for a righteous man will one die; yet perhaps for a good man someone would even dare to die. But God demonstrates His own love toward us, in that while we were still sinners, Christ died for us" (Romans 5:7, 8).

## Dirt and Dust

When God cursed the serpent for its part in Adam's sin, it was at once a curse on both the tool used to cause Adam to stumble and fall, and against the demonic intelligence behind it. This is why Satan has the name Old Serpent (Genesis 3:14) and great Red Dragon (Revelation 20:2), for he was from the beginning a liar (John 8:44) and murderer of the human race. God so esteemed the value of man that He cursed both the instrument and the one who wielded it in wickedness. He humbled both Satan and the talking beast and laid them low in the dust.

As a subordinate creature the serpent was one that had previously paraded before Adam. Adam had considered it a gentle and good creature, for so it was, and named it along with the other beasts. Yet it became abused in a terrible way, like a jeweled sword snatched from the hand of a faithful servant and made to kill the king's son. It was good before but became degraded through murderous designs and so came to suffer the vengeance of an angry father. Now, far from being good, it was by God's curse reduced to a vile and humiliating rank; it was to grovel in the dust and lick it up: "you shall eat dust all the days of your life" (Genesis 3:14). "Just as a loving father when punishing the murderer of his son, might snap in two the sword or dagger with which the murder had been committed,"[246] wrote Chrysostom (347–407), ancient Bishop of Constantinople, so did God spoil the serpent.

The Green Dragon glories in reversing good to evil and prefers to drag man through the dirt and dust, speaking of mankind in general as something low, as if he were less than a serpent. To eat dust is the sign of a base and sordid nature, and man is a contemptible creature. Christians and environmentalists can agree on this. But he is also God's image, in spite of all his vileness. To the Green Dragon this belief is unacceptable hubris.

For man alone to be the image of God is intolerable. If he is God's image at all, then he is no more so than a virus or an atom. If we continue to insist on being the lone image, then environmentalists seem to need to satisfy the urge to make plain that we are no more than animals. How different is the realism and balance of the Bible, which urges urgent consideration of our place in the world. Blaise Pascal wrote, "The more enlightened we are, the more we discover greatness and misery in man."[247] Both aspects of our existence are plain. Man is the image of God, but disfigured and deformed. As Jonathan Edwards, the puritan president of Princeton, wrote, "The serpent that tempted man was a talking, reasoning beast, and an incarnate devil; and such is man now become."[248]

## Blood and Mud

To many environmentalists man is an automaton, a clump of blood and mud. Not even Plato's silly ghost in the machine receives much love from some environmentalists. To others the problem is that the machine, the flesh, is not properly reverenced as holy in itself, and they accuse Christians of failure to understand their proper relationship to nature. Al Gore, for instance, writes of Western civilization: "We have misunderstood who we are, how we relate to our place in the creation."[249] Gore is correct insofar as his critique focuses on the vapidity of abstract thought unhinged from spiritual foundations. Environmentalists, like Gore, often cite Plato and other Greek influences as a cause of this unfortunate misunderstanding, and to this we will turn in detail in a later chapter.

Imperious human pride is unbecoming of a creature formed from dust of the Earth. Gore's solution, however, is to reach back further, to the ages of supposedly Earth-focused religion. Platonic philosophy, he argues, "assured that later explanations of the workings of the world would become progressively more abstract."[250] We must, he argues, respiritualize our low view of the physical. Jesus is not, however, a product of Greek philosophy, so going beyond the Greeks to the animists is futile. Going back to Jesus, to His Word, is what everyone needs.

Modern Christendom receives just criticism for its unhealthy emphasis away from God and towards mystical feeling and emotionalism. An excess of abstract and rational intellect is hardly the problem. The problem lies partly in an explicit rejection of intellect.

Many will correctly admit that the image of God is a spiritual quality of man. It is not a physical quality in the sense that God looks like the form of a human at all. But they will take it further, having received encouragement from the narcissism of modern civilization, to search for God by looking inside. Such is the essence of humanism. Columbia University President Lee C. Bollinger has expressed clearly the emphasis of this worldly wisdom. "Above all," he says, "we should want the most vital expression in the arts because of what it can say to us about ourselves."[251] In short, if man contains the image of God within the vessel of his body, then each one of us has the potential to discover God in ourselves, the divine spark. That is not the Christian way.

It is the spirit, not the flesh or biology, which sets man apart from the beast. But it is wrong to think that it is the spirit alone that is exalted, and not the body, that it is only by the spirit that we can realize unity with God, and therefore with other men. One can sympathize with the weakness that leads to such faulty human logic, but one must contradict the idea. The key mistake in this argument is that man can discover the image of God within himself, just as Gore and others argue we can find it elsewhere in nature. God's Word teaches that man does not *contain* the image of God as some mystical spiritual quality, but that man *is* the image of God. It is not a brand or stamp that is the mark of God; he is the image of God, body and soul.

In the Biblical Christian view the physical body is therefore never denigrated or looked down upon but raised to the heights of glory.

It is not equivalent to the spirit, for the body must decay and suffer corruption because of the curse that lies heavily on all God's creatures, while the spirit never dies. But recognition of the inferior status of the body is no excuse to fall on our swords and pretend matter is unimportant; more about this later.

We are creatures with physical and spiritual natures mysteriously united into a whole. The physical nature, like the spiritual, is fallen, and the body is subject to the curse of death, but the body will rise again. Jesus Christ will redeem the body, and He will glorify it. Biblical Christianity is not hyperspiritual and does not overemphasize the spiritual part of our existence, so critics should lay that tired canard to rest.

## Grace and Glory

In the Biblical view lies recognition of both the humility and the grandeur of man's being. As to his flesh, the origins of man are humble, for he was made from the dust of the Earth. However, even in the misery of his fallen state, it is faint praise to call mankind great and marvelous, for the finger of God made them, male and female, and set them apart. God gives us special care and attention, and incredibly sets His own heart on us. There is a natural tension in both the disgrace and the glory that are part of our condition. So the sweet psalmist sang in wonder, "What is man that You are mindful of him, and the son of man that You visit him? For You have made him a little lower than the angels, and You have crowned him with glory and honor" (Psalm 8:4–5).

Even in our fallen state mankind remains the image of God. And though much reduced our native glory is still exquisite. Holy angels in heaven, mighty spirits who always see the face of God, who do His will, wonder at the grace of God to mankind. We confess to being close of kin to angels and demons. But at the same time in a certain way God places man on a level with animals. It is not that man is an animal, far from it. Yet like the beast fallen man is subject to sickness and death, his body reduced once again to dust (Ecclesiastes 3:16–22). Through the fall mankind received a frail and mortal nature, a low, animal-like estate, which should forever keep him aware of his need for humility.

How the angelic beings all must have puzzled at the special care God had for man, how delicately He ordered their lives, and how He made the great angels minister to them. Finally, at the coming of Christ, angels went to school and learned the reason. More than ever before they learned the wisdom of God in the display of what had previously been to them the great "mystery" of Christ (Ephesians 3:9, 10); this is the glory of the gospel.

Man's magnificence and honor is even greater than that of the holy angels, though they are mightier than we. Twenty-four elders, seated in honor, clothed with crowns of gold and white robes, surround God's throne in heaven (Revelation 4:4). These elders represent Christ's church, made up of redeemed mankind, for they sing to the Lamb of God: "For You were slain, and have redeemed us to God by Your blood out of every tribe and tongue and people and nation" (5:9). Thus the Church of Christ, made up of redeemed mankind, worships directly before the throne of God in heaven.

1 Corinthians 11 shows how angels gather and watch the worship of the saints on Earth. Perhaps they even join in as we draw together to the throne of God. If so they, who are no longer ignorant of God's great care for man, must count it a great honor. Chrysostom, an early hero of the church, indeed suggests the angels, though they are unseen, actively participate in our public worship of God: "Thou standest with angels; thou singest with them."[252]

In this sense Christian public worship is the glorious communion of God with man and angels. As if our privilege and glory were not enough, the greatness of man is magnified, for what an honor it is to participate in so holy an act with such noble creatures, and even to sing with them the songs indited by the Holy Spirit! Though the humblest Christian, as a son of God, has a greater inheritance than the mightiest angel, yet he is still to come into the full enjoyment of that inheritance.

And man's glory reaches still higher. In heaven the angels cluster around God, eager to obey Him and bask in His presence. Revelation 4:4 says that our human representatives also surround the throne of God: "Around the throne were twenty-four thrones, and on the thrones I saw twenty-four elders sitting, clothed in white robes; and they had crowns of gold on their heads." But the saints in heaven enjoy what Jonathan Edwards called "a superiority of privilege"[253] since the twenty-four elders, who represent the church and surround the throne, find angels surrounding them. Though the angels crowd close to God, they are never so close as redeemed men. The angels stand round about the elders as the elders stand "round about the throne" of God (7:11). What astonishing glory God gives the saints to have assigned to them a seat closer to God than the potent host of angels! Christians should never forget that the living God calls them, male and female, His treasured sons (Romans 9:26).

## Conclusion

There is a chair in the department of philosophy at the University of Edinburgh in Scotland. On it lays an inscription with the following words, set there at the behest of Sir William Hamilton: "The greatest thing in the world is man; and the greatest thing in man is mind." With such warm and true praise a sinful and proud man might become puffed up like an iguana displaying itself before the blazing sun. Thoughtful, Biblical balance is a definite need.

In the entire world, no one so wise as Solomon has ever thought about these kinds of issues. After a long and exhaustive study, here is his conclusion: "Truly, this only I have found: that God made man upright, but they have sought out many schemes" (Ecclesiastes 7:29). We have seen that God created man in His image as a flawless creature, but that he fell from this original state of knowledge, righteousness, and holiness.

Biblical balance requires recognition of both the original nobility and the later dishonor of Adam and his posterity. Biblical balance also allows us to cut through the web of deceit spun by the modern environmentalist movement; man is not so low as environmentalists would have him. What Blaise Pascal wrote is true: "Christianity is strange. It orders man to acknowledge that he is evil, even abominable. Yet it also bids him to desire to be like God. Without such a counterpoise, this dignity would make him horribly vain, or promote such humiliation that would make him terribly abject."[254]

Like a child that would rather throw its toys under the crushing wheels of a train than share them with another, ecotheologians and other environmentalists seem to think that righteousness is relative, and will argue pertly that if nature and the Earth are not sacred, then nothing is. This is a lie. The Bible says (Leviticus 24:21), "whoever kills an animal shall restore it; but whoever kills a man shall be put to death." Human life is more precious than any other creature, because it is human.

Princeton philosopher Peter Singer disagrees. He believes that "[t]he notion that human life is sacred just because it's human is medieval."[255] This, too, is the capricious lie of a brat who argues for whatever he likes, and against whatever does not appeal to him, at the moment. Singer might as well write a law forbidding the sun from rising tomorrow morning because he wants to sleep in. Whether we like it or not human life is sacred and infinitely precious because made in the image of God. There is no better reason. From Scripture we learn, "Whoever sheds man's blood, by man his blood shall be shed; for in the image of God He made man." (Genesis 9:6). In the end, image is everything.

# CHAPTER FOUR

## Naked Ape: How Nature is Not Divine, and Humans Are Spiritually Superior to All Other Creatures

*Behold the iguana puffing itself out to make itself a man!*

South African Proverb

## Mountain Climbing

Mountains such as K2 on the border between China and Pakistan reach to scrape the roof of the world like a jagged tooth against the sky. Unlike Everest, which some modern aficionados claim has become a bit of a tourist trap, the challenge of a peak like K2 remains unquestioned. It is hardly possible to "dumb down" K2; without oxygen it is virtually unassailable, and only a handful of mountaineers have succeeded in mounting the peak; in an August 2008 attempt 11 died trying.[256] One out of every four people that have tried to climb K2 has perished in the effort.

Were the peak severed from its shoulders, those broad glacial valley bottoms near its roots, and dumped on the beach it would stand around 10,000 ft, which means that fit casual hikers could handle it, barely. Although at sea level this alone would be no challenge for serious mountaineers, at altitude it is a death trap known as the Savage Mountain. No animal is foolish enough to struggle to climb this stupendous peak, yet humans routinely die trying. This speaks directly to the peculiarities and glories of the human spirit, subjects we will address in this chapter.

Why try to mount a peak that stands on top of the world? A *New York Times* correspondent had this to say, following the 1924 disappearance on Everest of George Leigh-Mallory, the great English mountaineer: "It is the necessity of human nature that practical records should be broken; and to affect the inability to understand why men are irresistibly attracted to the chance of doing something first or going somewhere first, is a waste of time. They are built that way, and to the very impulse that rules them the world is enormously indebted."[257]

This is precisely the language and motive that upsets radical environmentalists. Perhaps surprising, then, is the radical environmentalist love affair with mountains. Thoreau, Muir, Leopold, Næss, and others climbed mountains with relish.[258] In doing so, they believed they embraced the sacred mountain on her terms, resting their bodies close as on the bosom of Earth Mother. So it is not mountain climbing *per se* that they hate, but wanton defilement of mountains inflicted by the feet of men lacking proper reverence for the rocks. Just like mountaineers out to master mountains, the forces of nature, and perhaps their own fears, so humans hungrily seek to scale other peaks, and scorn ancient natural limits.

Rather than seeing glory in mankind's constant search for new horizons, environmentalists see only barbarians out to defile mountain peaks; they want desperately to pull mountaineers down from those lofty heights. Just who do these puny humans think they are? Humans clearly have an overgrown "civilization complex,"[259] need to learn their place, and stop being, well, so human. It is hubris, then, that is man's original sin, not eating the forbidden fruit. Feminist ecotheologians like Roman Catholic Matthew Fox say that hopeless anthropocentrism is exceeded in wickedness only by arrogant androcentrism, the view that males are central in everything. In Fox's story of the fall, echoed by many environmentalists, a few millennia ago the age of benevolent female rule ended, along with its mythical nurturing relationship towards Earth. Matriarchy was just another mountain mounted roughly by proud patriarchy.[260]

Christopher Manes, author of *Green Rage*, writes: "Charles Darwin's *Origin of Species* had already undercut the metaphysical underpinnings of anthropocentrism by displacing the notion of the *scala naturae*, the Great Chain of Being, which situated mankind in a privileged station above the 'lower' life-forms of a divinely instituted world. Evolutionary theory denies the existence of a hierarchy of beings, declaring that there is only genealogy, similarities and differences arising out of a three-and-a-half-billion-year saga of organic inheritance in which we are only minor players. Taken seriously, evolution means there is no basis for seeing humans as more advanced or developed than any other species. Homo sapiens is not the goal of evolution, for as near as we can tell evolution has no telos—it simply unfolds, life-form after life-form. Elephants are no more developed than toadstools, fish are no less advanced than birds, cabbages have as much ecological status as kings. Darwin invited humanity to face the fact that the observation of nature has revealed not one scrap of evidence that humankind is superior or special, or even particularly more interesting than, say, lichen."[261] Let us examine these ideas in this chapter.

## Bad Karma

As we have seen, it would be an understatement to say that environmentalists do not approve of the Judeo-Christian idea that man is somehow privileged. Our race bears no special grand place in the world order, let alone as the image of God. When Al Gore calls for an ecocentric view to be the central organizing principle of a new society, he is calling for no more than Manes, and no less than Nietzsche, who wrote: "We no longer [after Darwin] derive man from 'spirit,' from 'godhead,' we have put him back among animals.... He is by no means a crown of creation; every being along with him is at an equal stage of perfection."[262] If this is true then every living being is also the image of God.

It is to this idea that we now turn our gaze. While at first it may sound noble and fair-minded, such a view seems more concerned in practice to raze man than raise nature. Because of its lying heart, it succeeds only in demolishing the value of both, making them lower than dirt. Since a cancer cell is no less valuable than a man, a man is no more valuable than a cancer cell. In such a world, one should not be upset if a neighbor pejoratively styles him a female dog. It might have been intended, after all, as a compliment. Hyperventilating over such an interspecies slight is hopelessly anthropocentric.

So we descend in a chaotic jangle to the inexorable view that human morality is a pathetic farce; all that remains is power, or perhaps a supposed need. So-called 'deep' ecologist Næss explains: "[I]t is against my intuition of unity to say 'I can kill you because I am more valuable' but not against my intuition to say 'I will kill you because I am hungry.' In the latter case, there would be an implicit regret."[263] Understand therefore that if an arch environmentalist like Næss ever becomes so philosophically consistent that he finds it necessary to kill you or your children to save a stand of trees, or because he is both famished and more powerful than you, it will never be personal. He might even feel some sorrow, great heart that he is.

Is it worse to cut down a tree or to murder a human? To a philosophically conscious and consistent environmentalist the answer clearly is that it depends. Cannibalism is not necessarily bad, it merely depends on whether the strong are hungry or not. The late Carl Amery, a Roman Catholic philosopher who helped set up the German Green Party, summed it up like this: "We, in the Green Movement, aspire to a cultural model in which the killing of a forest will be considered more contemptible and more criminal than the sale of 6-year-old children to Asian brothels."[264]

## Nature, Man, and Things

Environmentalists claim to reverence life, but their overwhelming passion appears to be hatred for man, most of all for his overweening self-importance. When humans consider their lives, or the lives of their children, to be more important or valuable than, say, a mosquito or a tree, then to the environmentalist they are engaging in a subjective ethic. To do so is anathema to the environmentalist, and that virtually all humans do so explains, at least partially, why the leaders of the Green movement, those big mouthpieces for the Green Dragon, make such rude comments about humans.

Albert Schweitzer, the humanist famous for his medical work in Africa, summarizes further the environmentalist ethos: "The ethics of reverence for life makes no distinction between higher and lower, more precious and less precious lives. It has good reasons for this omission. For what are we doing, when we establish hard and fast gradations in value between living organisms, but judging them in relation to ourselves, by whether they seem to stand closer to us or farther from us. This is a wholly subjective standard. How can we know what importance other living organisms have in themselves and in terms of the universe?"[265] How can we know, indeed? If we cast off the objectivity of the Bible, then we are adrift in a sea of human subjectivity.

Schweitzer goes beyond the usual hackneyed criticisms of Christianity and challenges fellow humanists and environmentalists. In the Darwinian view of life, man is the pinnacle of evolution, and primates are our idiot cousins, but cousins nonetheless. Social Darwinism, as practiced and preached by the humanists and eugenicists of the last century, acted on the persuasion that mankind was the center of value on planet Earth; all comrades are equal you see, but some comrades are more equal than others.[266] Schweitzer argues, as have so many in the environmentalist movement, that such anthropocentrism is an unsustainable arrogance. David Ehrenfeld in *The Arrogance of Humanism*[267] thinks it touchingly naive for humans to imagine they are somehow the demigods of evolution.

But if man is not the evolved peak of the natural order, then what is he? He is a rat or a pig or a dog. He is a cancer cell. And if a cancer cell is not divine, then neither is man. It is just a cancer cell. Professing Christians who are attracted to environmentalism need to come to terms with the misanthropy inherent in the Green movement. Environmentalist sensibilities may be broad, and include reasonable folks interested in clean water and responsible human behavior, but when its diffuse principles settle, like the dust of nuclear fallout, the answer that invariably radiates from the movement is that mankind is insignificant and unimportant—and irremediably wicked.

## Body Lice

If man is no different from an animal, why do humans wear clothing, and animals not so much? All over the world people wear clothing, whether in the jungles of Borneo or of New York City. Wearing clothing seems to be as common a practice as eating. One's answer can help to discover quickly which camp one prefers, for the answer given by God's Word opposes the Green Dragon.

Not to put too much of a point on delicacy, followers of the Dragon, and not a few professed followers of the true God, resonate with the argument that humans wear clothing because of body lice. Well, really because of the absence of lice. According to this just-so tale, told so well by Nicholas Wade in *Before the Dawn*,[268] the human body louse lost its ecological niche when our Neanderthal ancestors lost ape-like body hair, surviving on as the head louse. In their modern form body lice cling to fabric rather than onto hair and skin, and they have slightly different genetics than head lice. They ostensibly evolved only when our wearing of clothing gave them a new ecological niche as a fashionista. One can guesstimate the time to the common ancestor from studies of the differences in head and body lice. Further layers of assumptions lead to the conclusion that humans began to wear clothing around 78,000 years ago, just at the time that clothing began to wear lice.

Some have argued the origin of clothing was functional and decorative. The naked ape was cold and covered itself with decaying plant or animal matter as protection against the environment. Or maybe the design of clothing was to draw attention to the wearer and display the body's beauty and sexual utility. Such explanations are the outer layers of a philosophical onion that do not reach to the heart of the matter. Speculation has its place, but we may be grateful that is not necessary here, since the Bible gives the true history. And is it not better and reasonable to draw our lessons from the "more sure word of prophecy?" God records who, what, when, where, and how—He was there, and He deigns to tell us His observations regarding the first clothing. In one word, the reason humans wear clothing is sin. However, sin is not an idea familiar to your average louse, or cancer cell for that matter, and it seems rather unfair to burden the poor little beasts with such enduring debts.

## Heart of Darkness

Why do environmentalist ideas seduce so many people when the movement's foundations are, root through fruit, opposed to the Biblical view? Given the announced (though not practiced) austerity of the movement with its tiresome scolding, incessant recycling, cold showers, and high fiber diet, at first it seems odd that the Western World, with its massive pleasure-seeker population and rapidly vanishing residual Christian sensibility, should find environmentalism compelling. The *Asia Times* pundit called Spengler[269] argues that narcissism inherent to the prevailing world spirit explains, partially at least, the love affair with shallow postures that allow for good photo opportunities. Narcissists like it when others watch their good deeds. Receding tides of liberty plus the increasing bonds of government control smudge boundaries between role-playing and coerced conformity, on the one hand, and indoctrination and deep devotion to the Green Dragon, on the other.

Romans 1:21–23 delivers a prime answer: "because, although they knew God, they did not glorify Him as God, nor were thankful, but became futile in their thoughts, and their foolish hearts were darkened. Professing to be wise, they became fools, and changed the glory of the incorruptible God into an image made like corruptible man—and birds and four-footed animals and creeping things." A cancer cell is simply doing what Næss says it must do; it is hungry, so it eats; it is not immoral. And since man is a cancer cell too, he cannot be immoral, certainly not in the sense of having an eternal obligation to Jesus. And if man is not immoral then sin, like other religious connotation words, is a wax nose to mock, pinch and twist. *God* is then a word made from silly putty. Why then, perhaps the Green Dragon is God. When we begin to grow a philosophy of life in the irrational axiom that man is but a material creature, and deny his spiritual qualities, the natural result is to make of him an evolutionary animal red in tooth and claw.

Atheism is so absurd and contrary to reason that it is intolerable for many who will nevertheless not surrender and confess Jesus as the one true and living God. Here lies the danger of offering a mirror to a blind man—show them the foolishness of one place and they simply move on to fall into a different ditch. They take the road of compromise, a way of escape, to see God as a spiritual light, a kind of luminous vapor or invisible ether that permeates everything. Christ's religion is not agreeable to such depraved pagan ideas; His Apostles attacked paganism head-on. "God, who made the world and everything in it," said Paul the Apostle, "since He is Lord of heaven and Earth, does not dwell in temples made with hands. Nor is He worshiped with men's hands, as though He needed anything, since He gives to all life, breath, and all things" (Acts 17:24–25). Because He made all, it is absurd to think that one could condense or diffuse His essence into the few atoms of an image made of wood, stone, or any part of nature. Nature reveals His glory, in part, like the glory of the moon as it reflects the greatness of the sun. However, the sun would consume the moon if it came too close.

Pagans recognize glory in the cosmos. Even the most hardened atheists have written glorious paeans to the beauty, even the divinity, of nature. Carl Sagan, the late physicist from Cornell University, wrote, "Our feeblest contemplations of the Cosmos stir us—there is a tingling in the spine, a catch in the voice, a faint sensation, as if a distant memory, of falling from a height. We know we are approaching the greatest of mysteries."[270] It is interesting, and perhaps ironic, that a Sagan is a deputy of the Jewish High Priest. A Sagan is the second highest functionary of the Temple.[271] Until he went to meet his Maker, our own Sagan, Carl, fastidiously performed the dramatic role of a deputy high priest of a new mythology. Like Sagan, even if left alone on an island of godless thought, the atheist fool cannot escape the warmth of his human heritage and still wants to say of the golden sunrise, or the boundless starry night, "Isn't Mother Nature wonderful, even divine?"

Such flashes reflect the drive to worship, an irresistible impulse built into human hearts by their Creator, and it will find a way out. God's revelation surrounds mankind. It speaks eloquently to God's power and glory. So why is it that atheist groups roam the land, seeking to break down all remembrances and memorials to God? Here a statue; there a plaque; over there the Ten Commandments. In America the written Revelation, the words of the Bible still relatively common in the public square, is purged from view in an obsessive manner reminiscent of Lady Macbeth's damn'd spot. They hope once they have cleansed the land the still, quiet voice that disturbs their peace will be silent. It will not, for even atheists are made in the image of God. Though the atheist lives in the flesh, his innermost nature echoes eternity. As hard as it is to tear the flickering stars from the sky, so is the urge to worship etched indelibly into their souls (Ecclesiastes 3:10–11).

The reality is that, though popular a generation ago, atheism is going out of fashion. God is not dead, but atheistic secular humanism is an endangered species. God created humans with an inherent need to worship, so atheism is unsustainable. It only exists as a serious philosophy through violence, as in socialist utopias like North Korea.

Now if men will not worship God, they will worship what God has made, for worship they must. Nature worship is pure paganism. It has enjoyed, for some decades, a rebirth in Western Civilization as shown by the exploding pop genre of Wicca and other nature religions. It is increasingly popular as a subject in countless Hollywood movies and TV shows, for example *Charmed*, *Buffy the Vampire Slayer* or the *Harry Potter* dynasty.[272] James Cameron's blockbuster, *Avatar*, has Americans as evildoers and tree worshippers as the good guys. Says Cameron, "Look, at this point I'm less interested in making money for the movie and more interested in saving the world."[273] Pagan spirituality hankers after the day in which spirits of the river, leaf, and woods once again permeate nature.

Anti-intellectual streams in modern society resonate powerfully with the uptick of interest in occult powers and paganism. Most Americans claim to be Christians, and nine in ten believe in the spiritual dimension, but most also believe that what another person says is truth is merely an opinion. Charles Spurgeon met the measure of such arguments. He said, "He who turns a morsel over and over in his mouth does so because he likes the flavor, and he who meditates on evil loves it."[274]

So the church faces a revival of societal paganism, a new Gnosticism seeks to infiltrate the professing church. In many cases various aspects of environmentalism are the vehicle.

Those not enthralled by the pagan renaissance may nonetheless be appreciative of its various rituals that enshrine the energies and rhythms of nature. In 1990 a group of notable scientists, many of whom claimed atheism as their religion, called for a pagan reinterpretation and fusion of all religions. In a statement, authored by the prolific Carl Sagan, the signees agreed, “As Scientists, many of us have had profound personal experiences of awe and reverence before the universe. We understand that what is regarded as sacred is more likely to be treated with care and respect. Our planetary home should be so regarded. Efforts to safeguard and cherish the environment should be infused with a vision of the sacred.”[275] This was one of the first such “sacred” statements issued by a distinguished group of atheist scientists, but not the last.

One of the most astonishing observations associated with the unholy coupling of environmentalist confusions and presumptions is the applause it has received from professed Christians. All of Sagan’s words, which I have quoted, could easily have slipped from the lips of Christian leaders. Of course, Christians are those who recognize the existence of the sacred, and treat the idea not merely as a god-word to stroke like a gorgeous ginger pussycat. Christians recognize that spirits good and evil exist and play a role in our lives. Yet the modern overemphasis on feeling and experience in the church has made Christians susceptible to false ideas about the sacred, as if the narrow way taught by Christ is to reduce God’s essence into a leaf, or a river, or some other created thing such as the eyes of a child. Historian Herbert Schlossberg compellingly argues, in *Idols for Destruction*,[276] that the idols of the nations against which Israel fought were precisely of this kind—aspects of nature (as in pantheism), or idols of history and creation (as in evolutionary theory).

## A Sacred Earth Charter

In our day, the *Earth Charter* codifies the sacredness of the Earth. It forms the Green replacement for the Bible’s Ten Commandments. Steven Rockefeller, head of the *Earth Charter Commission* (incidentally the brainchild of Maurice Strong) and Professor Emeritus of Religion at Middlebury College in Vermont, coordinated the drafting of this deeply religious, but certainly not Christian, document. It uses connotation language that to gullible Christians might sound laudable. It argues that protection of the Earth "is a sacred trust," that humans must identify themselves "with the whole Earth community," and that we should "live with reverence for the mystery of being, gratitude for the gift of life, and humility regarding the human place in nature."[277]

Not only do many atheists, pantheists, and professing Christians support these ideas but also the 1,000-member US Conference of Mayors and other politically influential entities. Buddhists and Baha'i do, too, and Jews through Jains. Influential twentieth century Jewish theologian Abraham Heschel, who died before the real explosion of Green political power, nevertheless argued for some of the same foundational principles, such as fitting reverence[278] for the Earth and all that exists. "Certainly," wrote Heschel, "just to be is a blessing, and to live is holy."[279] But in a world of sinners, when we make *everything* sacred or holy, therefore worthy of reverence, *nothing* is. This is reminiscent of Oprah Winfrey's statements quoted in the conclusion of Chapter 2, echoes Al Gore and the ecotheologians, and argues for the evolutionary convergence of religions that so many in the environmentalist movement fantasize about. This view that ordinary existence is holy in itself has become popular in all major religions. It is a pillar of the world spirit.

Pantheist innovator David Spangler agrees: "If something is sacred, it is assumed to have value beyond its form, its usefulness, its duration, and its products. It is valuable and precious in and for itself. It is worthy of respect and honor, love and compassion; it is worth entering into communion with it. Its very being is its only justification, it needs no other."[280] Rosemary Ruether espouses a not so different view of nature-human kinship, for instance: "A profound spirituality would arise if we would attempt to experience this kinship and make it present in our consciousness."[281] Here is a prominent professing Christian theologian, and a feminist to boot, who finds no problem with the idea of Earth sanctity, imputing to it intrinsic value, and so making common cause with Green atheists and pantheists. Indeed, she believes this idea is Christian and comes to it through the path of evolutionary epochs.

No wonder an arch atheist like E. O. Wilson is so positive about the prospects of what he terms the Green evolution of Christianity. Whether they like it or not those in the professing church who argue for resacralization of nature align themselves with policies whose end is the crushing of mankind, and the exaltation of nature and nature's gods. It bears repeating that the fond desire to exalt nature, however well intentioned, has the same effect as calling a man an animal.

When speaking of the need for the church to be environmentally conscious it is notable that the words of theologians enjoy extensive continuity with the repertoire of the Green Dragon, liberally sprinkled with god-words. Even evangelical Christians receive routine urgings, for instance, to see "Creation itself as a collaborator"[282] and to appreciate the "inherent value of the earth and its creatures."[283] How smoothly such ideas are insinuated into the core of modern Christendom. *Christianity Today*, the "Magazine of Evangelical Conviction," routinely runs articles praising the virtues of "Creation Care" and other environmentalist shibboleths. An editorial explained how "The Bible is not an enemy of the environmental cause, but its greatest asset."[284] *Eerdmans*, a major evangelical publisher, now regularly publishes tomes on ecotheology.

## Into Philistine Hands

Environmentalism and its Christianized incarnations impute to nature the idea of intrinsic value. Methodist minister Vernon Visick argues that Christ practiced and preached such an environmentalist reading of our relation to nature.[285] But is this the case? Is the Bible environmentalism's greatest asset, and does God ascribe intrinsic value to nature? To answer these questions it should be clear that one must turn to the Bible. In the previous chapter, we recalled the time when Moses stood on holy ground before the burning bush, and we showed there how holiness is not an inherent quality of matter that fills a space. Only God is intrinsically holy, and anything else is holy only because He has so designated it. God's holiness is intrinsic; anything else's holiness is imputed. Holiness is a spiritual quality that presses material qualities to the side, simultaneously engaging the mind and imagination.

God often condescends to our finite and frail nature, and allows physical metaphors to paint spiritual realities, such as holiness, on the canvas of our imagination. For instance, His mercy is as high as the heavens above the Earth (Psalm 103:11). It seems odd that Bible-believing Christians suffer frequent accusations of being wooden literalists when, in fact, they have little difficulty in understanding these kinds of metaphors.

We know the altitude of the exosphere and it is not infinity, yet only an atheist literalist with an ax to grind might seriously believe this shows the Bible is in error about God's infinite mercy. One should not need to explain the intent of metaphors. They are word pictures that clarify hidden realities. It is only in Hollywood movies, not in the real world, that cats and dogs rain from the sky. Religious environmentalists are the truly wooden literalists. For instance, according to Brian Walsh, a theologian in the Christian Reformed Church, when the Bible speaks of trees singing (1 Chronicles 16:33) we should realize that this is beyond a poetic picture. Says he, seriously, "If we speak of metaphor only … we keep trees and humans in an I–It relationship; our attitude remains hopelessly anthropocentric."[286] He wants us to go beyond metaphor and genuinely listen to the trees in a respectful I–Thou relationship, channeling their earthwise messages to humanity.

Though professing Christian environmentalists make much of Scriptures that speak of skipping hills and groaning Earth, the only living stones that Christ speaks of are Christians. If Christians are living stones, no other stones receiving such distinction, then the other stones are not living. Peter is a living rock, all Christians are living stones in the heavenly courts, and those that overcome will be pillars in God's temple. By contrast mountains are not themselves spiritual; they are inanimate and without intrinsic spiritual qualities.

Jesus said, as He passed the bread, that it was His flesh and the wine was His blood. People thick as ten bricks or twisted as a knot became confused or offended at His words; early Christians were seriously accused of being cannibals.[287] Jesus said, "Most assuredly, I say to you, unless you eat the flesh of the Son of Man and drink His blood, you have no life in you" (John 6:53). Though Judas ate bread and wine at the Lord's Supper, later events made it clear that he never had spiritual life in him. He was spiritually dead for he had never been united to the Lifesaver through consumption of the spiritual blood and flesh of Christ. Metaphors are not reality. In no sense are stones and other inanimate objects spirit filled. Christians are. But why stop here? Once we grant spirit to trees and rocks, we can advance. Why not start a new physics in which we pause daily to listen to sayings of the atom? "Spirit of the atom, speak, we listen!"

In His holy worship, God provides tokens of His presence, sacred ensigns to serve as metaphors that allow His people to stretch heavenwards. A most prominent example of the sacred is the physical place of worship. God tabernacled among the Jews while they journeyed from slavery in Egypt to liberty in the Promised Land. He provided a tent of meeting, the Tabernacle, which He made a holy place. Everything else that God associated with His worship He also labeled as holy, from the people required to enter the holy place, to the clothing they wore, even to the sacrifices He obligated regular folk to bring.

In one representative instance, recorded in Exodus 30:36, God called on His people to provide special incense for worship, the recipe prescribed by divine authority. "And you shall beat some of it very fine," He says, "and put some of it before the Testimony in the tabernacle of meeting where I will meet with you. It shall be most holy to you." Even the incense was holy! But only a superstitious person would smell a sweet spice and think it held intrinsic holiness. Nothing in its atomic structure set it apart as holy; God alone declared it holy by setting it apart from ordinary use to a peculiar use in His worship. The perfumed incense was a physical metaphor for a spiritual reality and served as a symbol of the acceptable sacrifice and for prayers of the saints (Psalm 141:2).

Later, when God provided a fixed location for His worship, He was said to dwell there between the angelic cherubim (2 Samuel 6:2). Yet we know the stuff of space cannot actually contain God and that God does not live in temples made by human hands (Acts 17:24); the cosmos itself cannot contain God. Rather than imagining, as a materialist might, that one could somehow chain God to the Ark, the correct understanding is that the Ark was a visible token of the presence of God. Intrinsic holiness is not found in the material things but in God alone, who is a spirit—holiness is not to be sought elsewhere.

God even explains how the holy things we have mentioned are "a shadow of the good things to come" (Hebrews 10:1), not the actual things. Thus, the holy things on Earth that we might touch and taste are a foretaste to encourage our imaginations to linger on the eternal things of Christ's kingdom.

At the dedication of Solomon's Temple, as it became known, the king of Israel guarded against the temptation to imagine some intrinsic holiness to the rocks and stones and other things linked with it. "But will God indeed dwell on the earth?" asked Solomon, and he immediately answered his rhetorical question: "Behold, heaven and the heaven of heavens cannot contain You. How much less this temple which I have built!" (1 Kings 8:27). God allows no place for superstitious regard of created things, as if they had some sacred quality inherent in their nature. Acts 7:48 repeats Solomon's reminder, "However, the Most High does not dwell in temples made with hands." God is a spirit, and no material building can confine Him. Yet, for a time, He chose to associate and manifest His presence, in a tangible way, first with the Tabernacle in the wilderness, and later with the Temple in Jerusalem. It was His presence alone that made them holy (Revelation 22:3).

But the world is superstitious and chooses to pretend things are holy, when these are merely preferences based on prejudice and tradition. Cows are not holy just because a Hindu says so. Indeed, cows are tasty and good to eat because God says so. God's church also, when not led by His Word, often falls prey to some or other superstition of the Philistine kind.

On one ancient occasion, before the time of Solomon or David, Israel fought against the degenerate Philistines, and brought the Ark of the Covenant with them, as a talisman to help win the battle. Israel's leaders hoped to innervate her soldiers by its presence, to impart blessing to their warlike endeavors. The Philistines would also bring their idols into battle. Among the most important deities worshipped by the Philistines were Dagon, the male fish god, and Ashtoreth, the Great Mother or Queen of Heaven. Ashtoreth is given many other names, but one in particular is interesting. It turns out that Ashtoreth is that very Earth goddess that Al Gore speaks of so fondly. Ashtoreth, or Astarte, is the second name in an energy chant sometimes used in modern Wicca: "Isis, Astarte, Diana, Hecate, Demeter, Kali, Inanna."[288]

It is now not surprising to learn that Gaia, the Earth mother, is the name given her in the Cretan incarnation.[289] So the Jews were warring against those gentle organic Earth-focused pagan types one hears of so much in environmentalist circles.

The Ark of the Covenant contained the stone tablets on which God wrote the Ten Commandments, as well as Aaron's rod that budded and the manna from heaven. Two cherubim on the Ark's cover faced its center, their arched wings touching to form the base for the throne, or mercy seat, upon which God sat and judged His people. Unhappily, by their previous intercourse with the superstitious Philistines the Israelites had come to learn their heathen ways and, like them, to have superstitious regard for the physical things linked to worship. And what could be holier than the Ark? They vainly imagined the Ark, so centrally associated as it was with the Tabernacle worship, was infused with the spiritual vitality of God Himself. Without warrant or command from God, they presumed to pull out their version of Aladdin's magic lamp and brandish it like a sword before their enemies. In truth, they had replaced God's sovereignty with their own.

It would have been better if they had humbled themselves before God, rather than rely on magic tricks. God had told them how to fight, but rather than obey Him they preferred to try what had been successful for Moses and Joshua in glorious battles of yore. Instead of looking to obedience, they thought they had found the recipe for success; add one part God with one part Jew, stir vigorously on battlefield, and *voila*, one's enemies go up in smoke. They imagined the Ark tied God down, so while they dragged him along on their adventures, He would smite their enemies. And at first, the great symbol did have the desired effect—the Israelites cheered, and the Philistines paled. They knew full well Israel's great military exploits, and believed now that since the Jews had rubbed their magic lamp the genie of destruction would come flying out of it. "Woe to us!" they cried in horror, "Who will deliver us from the hand of these mighty gods? These are the gods who struck the Egyptians with all the plagues in the wilderness" (1 Samuel 4:8).

But God cursed His own people for thinking they could treat Him like a tame lion to show off before impressed crowds of Philistines at the Canaan zoo; instead of falling on the Philistines, the Israelites fell on their own swords; they suffered rout and the Ark fell into the hands of their enemies. God disciplined the apostasy of the Israelites because they trusted in the Ark instead of in Him, superstitiously imagining that His presence was inseparable from it—they treated God as the great Vending Machine in the Sky. Like their eco-friendly pagan neighbors, they succumbed to the false notion that matter had inherent transcendental qualities.

## Cooked Goose

In the New Testament, the overwhelming emphasis when dealing with holy things refers to their use in the worship of God. In general, plants or stones are not inherently holy, neither are people. But Christians, the people of God, are most certainly sacred, for God declares to them, "Do you not know that you are the temple of God and that the Spirit of God dwells in you? If anyone defiles the temple of God, God will destroy him. For the temple of God is holy, which temple you are" (1 Corinthians 3:16–17). Now, as has always been, holiness is associated with the presence of God. The walls of the Temple contained the Ark of the Covenant. Yet not even the stones of the Jerusalem Temple, where Jesus taught and worshipped, remain holy once God leaves. Jesus said that God would bring desolation on Jerusalem and the Temple. The Lord so carefully guarded His own holiness that when His people began to have superstitious ideas about the material means He provided for their training, His curse fell on these things, formerly declared holy, and He destroyed them.

The Jews of Christ's time, who were still called the people of God, and a holy nation, became cursed because they rejected God and His law. So great was His hatred of sin that God cursed his bride, Israel, when she blended His worship with idols. Then the holy Temple was cursed because of the unfaithfulness of the people of God. Rather than looking to Christ, they looked to the Temple, which had become for them an idol of history. God would see to it that even the stones of the Temple would be thrown down (Mark 13:2), said Jesus, without even a trace of nostalgia. The Jewish Temple is no more. We learn that it is the visible church of professing Christians, and their children, that are holy; Christians form the living stones of the new and eternal temple (1 Peter 2:5). So it turns out that rocks out of which the Temple was made are no more eternal than snowballs in Florida.

Some people object to God doing as He pleases and find it mightily offensive that He would curse rocks and wood. Famous for his many books attacking theism, particularly Christianity, Richard Dawkins, a professor at the University of Oxford, does not disappoint. A memorable passage in *The God Delusion* has Dawkins, tail coiled out of sight, launch with relish into a syncopated diatribe against his Maker. "The God of the Old Testament," writes he, "is arguably the most unpleasant character in all fiction: jealous and proud of it; a petty, unjust, unforgiving control-freak; a vindictive, bloodthirsty ethnic cleanser; a misogynistic, homophobic, racist, infanticidal, genocidal, filicidal, pestilential, megalomaniacal, sadomasochistic, capriciously malevolent bully."[290] Try gangsta-rapping that to Jingle Bells.

It is odd that Dawkins limits judgment to the Old Testament alone. He is either theologically ignorant or politically savvy, possibly both. God has not changed since the Old Testament, and does not change. If God were bad, anyone close to Dawkins should, for their own safety, gently shush him as a silly goose. But the God of both Testaments is, in fact, always good, which means Dawkins is fixing to be a cooked goose. Dawkins may as well rail against builders of bridges for making them so tall that people die when they try suicide off them. When we choose to disobey God's written word, we can hardly blame Him for our dish wish.

As in their failed adventure against the Philistines of old, many (but not all) of the professing Jews in Jesus' time, some 2000 years ago, had no intent to submit to Biblical standards of justice. They used the Word of God and the symbols of His presence (the Temple) to mask their cruel intentions with a superficial flavor of godliness. Sin may not linger in the presence of the holy God. Because He is just He does not ignore sin, but curses it and brings about judgment of it. Because He is powerful His judgment is fierce and His wrath terrible, not only on the sinner, but also on those things that are in a moral association to the sinner. One might think it odd for God to curse inanimate stones for the disobedience of some men, yet this is precisely the case. It happened in Genesis too, when the whole creation, which God had declared to be good, and therefore holy, became cursed because of Adam's sin, but much more about this later.

## Top of the Heap

The Biblical doctrine of creation asserts that the universe is not an everlasting mechanism but a temporary construction of the everlasting Father (Isaiah 9:6). "Nature herself," wrote warrior theologian Robert Dabney, "shows us the marks and proofs that she was not eternal nor self-existent."[291] Yet environmentalists, from the atheist end of the spectrum to those professing Christianity, ascribe eternal and holy qualities to the cosmos. Since this judgment is the prerogative of God alone, it is not for man to decide what is holy and what is divine; to do so is, if I may say it, arrogant anthropocentrism.

Man may be the image of God, but he is not God and dare not usurp God's place. That being the case, how can Christians possibly reconcile their service to God with service to a Green Dragon that by turns boldly declares the sacredness or divinity of nature, or softly uses language that implies that if nature is not sacred nothing is? Modern environmentalism has moved far from concern for proper stewardship of the Earth and Biblically based respect for God's creatures, towards a reverence for the Earth, simultaneously compatible with hard-core atheism[292] and spineless pantheism. Environmentalists' distorted vision of nature makes it the supreme context, or central organizing principle, for human concerns.

Broad evangelical Christian positions on environmentalism are approaching convergence with the liberal mainline and secular or pagan positions. What Al Gore wrote almost two decades ago is now almost conventional wisdom in much of the church: "Similarly I believe that the image of the Creator, which sometimes seems so faint in the tiny corner of creation each of us beholds, is nonetheless present in its entirety."[293] Gore is expressing a philosophy, known as panentheism, which holds that God is seen and known by His penetration of even the edges of the cosmos but not limited to or confined by it. This is the halfway house for pantheists who are irrationally fond of monotheism, a sort of limbo for those torn creatures one might call Christian pantheists.

God's Word teaches the precise opposite, that He is transcendent and apart from the cosmos. Man has intrinsic value because even the most corrupt man is the image of God, but the distant stars bear no such image. The tiniest corners of the cosmos display marks of the power and wisdom of God as Psalm 19 and other places teach. Glorious though it is, the cosmos is nevertheless not sacred and worthy of reverence or worship. It does not have inherent value apart from its relation to its Creator, and to Adam as the true image and glory of God. If we wish to know where our reverence should remain we need to go to God, and should not rely on the guesses of atheists, pantheists, and ecotheologians. To grant what is plain is not hubris. The cosmos is simply not in our league. Humans are of a higher order of creation than is the rest of the cosmos.

Even among mankind, we find heavenly distinctions. Though every man is the image of God, He declares only some holy. Among the fallen creatures of the cosmos, Christians alone are living and holy stones of the better temple: "Or do you not know that your body is the temple of the Holy Spirit who is in you" (1 Corinthians 6:19). Nature is simply not sacred. For Christians to adopt the pagan use of language is at best a gullible and sloppy misrepresentation of the truth.

Belief in the sacredness of nature boils down to a subtle disloyalty because it pretends reverence while it really insults, like Judas who betrayed God with a kiss. Sham respect for the things of God has been a merely formal camouflage for measures and decisions that oppose, embellish, or simply ignore the plain text of the Bible. What is happening in much of the church is that professing Christians, theologians and others, are simply employing the language of Christianity to wrap non-Christian and antichristian ideas, like a dash of explosive *plastique* beautifully gift wrapped but devoid of spiritual value. It is all flash and sizzle, and violence, but where is the beef?

## One with the Beef

"We admit that we are like apes," says evolutionary biologist Richard Dawkins, "but we seldom realise that we are apes."[294] This is deep. Dawkins admits to being a monkey's uncle, and is another fine representative of the current state of education in Western Civilization. One of his main observations in *The God Delusion* echoes Robert Pirsig in *Lila*, namely that "when one person suffers from a delusion it is called insanity. When many people suffer from a delusion it is called religion."[295]

As a good materialist Dawkins believes the present state of the cosmos is essentially normal when, in fact, the world is disordered and ruined in comparison to its former Edenic estate. This belief is one reason he flounders into evolutionary delusions. He believes in an innate association of man and cosmos because, according to his preferred mythology, man evolved out of inorganic atoms that slowly coalesced into life, and then moved through the multi-billion year spectrum of less to more complex animals. As atheist Carl Sagan said, "We are star stuff."[296] But then, pond scum is also star stuff, which was precisely the point. It sounds so generous to suggest people and stars are related, but in this case it morphs into more of a reproach for hoity-toity humans; the metaphor is not welcome.

Methodist environmentalist Bill McKibben, a professorial member of the Middlebury College eco-coven in Vermont, writes, "But the Earth and all its processes—the sun growing plants, flesh feeding on these plants, flesh decaying to nourish more plants, to name just one cycle—gives us some sense of a more enduring role."[297] Our flesh will grow old, decay and fade away, but the Earth and its rhythms are timeless, continuously involved in the womb of a fertile Venus of sorts. God's Word suggests the exact opposite is true.

It is the Word of the Lord that endures forever, but the Earth will wither like the grass and fade away (1 Peter 1:24–25). According to the environmentalist fantasy, when the nether realms of the Earth absorb our flesh we will engage again in the creative cosmic cycle of reduce, reuse, recycle and turn into the new improved evolutionary model. McKibben urges his readers to be aware of, and attuned to, these wonderful ancient rhythms of nature, rhythms with power to help one find fulfillment, a sense of meaning, and a place in the world. Perhaps, for the environmentalist, there is a certain symmetric nostalgia in believing that for the Earth to endure it must have already lasted forever. With heaven of the Bible relegated to a possibility, like Santa Claus, one may best grasp a sense of immortality by reaching for the stars.

"We are star stuff which has taken its destiny into its own hands,"[298] is the full quote by Sagan, from his award-winning, government-funded television documentary *Cosmos*. He is trying to make church bells peal in the ears of our hearts. His mechanical humanist vision of life, which finally speaks of humans as unimportant atomic specks on a pale blue dot adrift in the cold cosmic sea of space and time, does not compete well with McKibben's slightly warmer emphasis on Earth the nurturer. That is why the atheist Greens are irrelevant compared to the theist Greens who cling to the skirts of the kind mother figure that, though no less remote from the truth, imparts a more well-rounded sense of meaning and balance.

Atheism seems poised for extinction. It has no stamina, as former *über* atheist Jean-Paul Sartre discovered near the end, saying, "I do not feel that I am the product of chance, a speck of dust in the universe, but someone who was expected, prepared, prefigured. In short, a being whom only a Creator could put here: and this idea of a creating hand refers to God."[299] McKibben also revokes atheism in favor of a Creator, but he has so imbibed the model of the ancient evolved Earth that he blindly rejects other possibilities, and is as likely to catch truth this way as water in a sieve. He intimates that an almost preternatural satisfaction is ours to enjoy if we would but engage with things more ancient than we, and when we recognize in them the divine recreative work of God. But it is equally possible, and much less appetizing, to imagine a less pleasant description of the primal womb as a cosmic Venus Flytrap that repeatedly digests my flesh, even while I am alive and kicking.

Seeing he is a disciple of the idea of Darwinian evolution it is logical that McKibben sees grubby glory in the idea that we and the atoms and the beasts form a kinship that is nigh eternal. From this he offers comfort and a sense of immortality. So does the damned degraded doctrine of evolution, with its infernal ages, lead to rows of deathly dirty ditches that would not feed a frog. Darwinian evolution teaches how death and violence are natural, even good, and governed the animal kingdom hundreds of millions of years before *homo erectus* interfered.

As a professing Christian, he should be well aware of the incomparable comforts of the Rock of Ages, but as a professing environmentalist he prefers to hope in the comfortless coldness of the rocks of ages. One cannot help feeling some genuine grief for McKibben; he chews the bitter pill and does not even know it. Rather than long-lived the Bible teaches that the Earth we know was a recent creation, had a watery beginning, and will come to a flaming end. Evolutionary fairy tales drive much of the environmentalist movement into the tree of stupidity.

## Far Out Cows

When driving through the countryside one often meets pastoral scenes of cows patiently chewing the cud under the gay heavens, ruminating on the day's work, probably planning placid bovine adventures. Sagan and McKibben would have us ponder our cosmic role, a divine comedy, through enduring the timeless monotony of sports like cow watching. Not that one should bash cow communion—we can always learn something of spiritual value from the tangy smell of freshly dropped manure, just not much about our starry creation billions and billions of years ago out of the cosmic intestine. But how precisely is one to milk emotions of enduring wonder into lasting fraternal relations with the beef? Roughly fifteen hundred years ago Gautama Buddha claimed "[t]he cows are our friends."[300] Cartoonist Gary Larson of *The Far Side* fame draws cows taking tractors for joyrides, going on family holidays, and hiding their secret cocktail parties from humans. But the Green Dragon takes the farce to another level. It teaches the nativity of brother cow to credulous captive audiences at schools in a neighborhood near you, and no gainsaying allowed.

In one sense, a slender sense to be sure, it is true that we share something in common with cows, cancer, and Cassiopeia. In the Newtonian sense one might argue that our presence permeates all the physical order, since one can imagine a gravitational force field emanating from our bodies and extending ever so faintly to the hidden depths of the cosmos. When all gravitational fields from cosmic matter add up we find there is not enough gravity to hold everything together. Modern cosmology now engages in the search to explain why we cannot account for about 96% of our universe. "Dark matter" or "dark energy" are names given to explain those motions in the cosmos that contradict the present lore of physics. A kind of universal glue or primal matrix holds everything together, but it is a something that we cannot see, something immaterial and without mass in any conventional sense. Exotic is what physicists like Kip Thorne call this invisible theoretic mélange.[301] Christian pantheists have skipped in delight,[302] for they imagine they are seeing God face-to-face in the corners of creation, and that there is a fundamental unity to the cosmos imposed by a touch of divinity of which we and the cosmos share.

## The Cows of Eden

The book of Genesis argues against any such organic unity. Earth and its creatures form a part of the cosmos separate from other planets and stars, and separate from God and other spiritual principalities and powers. Scripture teaches an ordering of physical creation. Though by no means a continuum, as the scale of nature—the *scala naturae*—rises from atoms and stones there is an increase in the faculties of creatures, "growing still brighter and brighter," according to Puritan theologian Cotton Mather, "and more capacious, till we arrive to those noble ones which are found in the *Soul* of MAN; and yet MAN is, as one well expresses it, *but the Equator of the Universe*."[303] All humans are of greater value than animals, as Jesus makes plain in His rebuke to the Pharisees (Matthew 12:11–12).

God created angels as powerful, intelligent, immortal spirit beings separate from the material world, but able to interact with it. Each angel bears direct responsibility only for himself, as God created angels as individuals without any shared source. That is, there are no angel babies (Mark 12:25). When Satan and one-third of the angels fell with him in determined rebellion against God, each one sinned by his own choice and none was dragged down as co-dependent collateral damage. "We may say," writes theologian Frederick Leahy, "that angels constitute a company as distinguished from a race."[304]

It is the Scriptural teaching that angels are distinct from man and God and never come between man and God. In the case of demon possession of unbelievers, Scripture makes plain that the person of the demoniac is distinct from the beings controlling his body. For instance, Jesus and Paul addressed the demons present in the bodies of unbelievers as persons separate from the human host (Luke 4:35; Acts 16:16–18). And though they are more powerful than we, yet elect angels (1 Timothy 5:21) serve elect humans as "ministering spirits sent forth to minister for those who will inherit salvation" (Hebrews 1:14). There is clearly no unity between us and the spiritual beings called angels and demons.

Our planet was the first of the material things made *ex nihilo*, out of nothing, by the Word of God, certainly not from star stuff. The observant angels "shouted for joy" (Job 38:7) when they witnessed it. The order and timing of the physical creation shows that the Earth and everything else enjoys no direct connection with angelic beings. God tells us little about how He created the other planets and stellar objects. But this much we do know: they do not share the same formless and void state as the Earth (Genesis 1:2), whatever that may mean, and He seemingly created them *ex nihilo*. Interestingly, the "lights in the firmament" (Genesis 1:14) came into existence after the vegetation. It was a difference of just a single day, but the order is unmistakable; plants before the sun and other stars. Christian environmentalists want to have their cake and eat it. But how can the sun be billions of years in the making—the evolutionary story—while plants are waiting to photosynthesize its rays? My plants wilt faster than that. So neither the plants, nor man, created on the sixth day, share organic unity with the sun and other astronomical objects.

Since we share the same place in space, one might argue that plants, beasts, and man share a collective terrestrial consciousness. Such an argument drives Darwinian evolution a little closer to our imaginations. "If there is a relationship God has to us living beings," wrote eco-philosopher David Spangler, "then that relationship also extends to and includes the Earth. If we may participate in sacredness, then so does the Earth, not simply as a valued and cherished environment but as a fellow creature."[305]

Nevertheless, much as we cannot sing brother sun and sister moon with a straight face, neither can we appeal to geography or cosmography to allow us to sing of cousin cow or Mother Earth. The creation narrative indicates distinct creative acts for terrestrial vegetation (Genesis 1:11–12), marine life and birds (1:20–22), terrestrial animals (1:24–25), and humankind (1:26), all of them formed from inanimate matter, whether in the seas or on land. Contrary to evolutionary thought, God did not make sea creatures, whales for instance, related to humans but of a different physical origin. Where now is the supposed unity we share? If professing Christians would respond by rejecting evolutionary fairy tales they would partially defang the Green Dragon. For like fairies true Darwinian evolution has never been observed.

Christians can agree with environmentalists that Adam, as to his outward visible form, was made from the same stuff as the beasts of the Earth, and that Adam came into being after the beasts did. But then the whole bipartisan house of cards collapses. Modern environmentalism, clinging as it does to evolutionary mythology, teaches that more complex creatures arose from less complex creatures—simplicity at the start and vast complexity later. Yet God's Word teaches that the cows of Eden were fundamentally the same as the cows of today.

Apart from their terrestrial source, not much more light is shed on the creation of the animals. They are not star staff any more than we are; we are earth stuff. But our mode of creation is significantly different from the animals. God explicitly states the unity of the human race as all humans living today, male and female, spring from the same origin. All human life is fundamentally one, as the Bible teaches throughout, but not fundamentally one with all life. From Genesis 2:22 the record states that each human descends from the one man and one woman with no commingling of man and animal.

Our species springs entirely from our first parents. Adam came directly from the dust of the Earth and the breath of God. His wife and suitable helper God built from his rib, and then all their posterity proceeded from their sexual union. "Thus, we perceive," wrote theologian J. G. Murphy, "that the entire race, even the very first mother of it, has its essential unit and representative in the first man."[306]

Differences between the races, the most obvious of which is skin color, have led many to question the Biblical record. The ancient Athenians nursed the view that their blood remained untainted by inferior stock. Not to be outdone, and seemingly incapable of learning lessons from the past, modern Malthusian eugenicists such as the socialist Margaret Sanger, of *Planned Parenthood*[307] infamy, have an evolutionary vision of an America and a world that refuses birth to "weaklings" and "the unfit,"[308] racially defined. Paul the Apostle has reduced to ashes all such pride by teaching that God has "made from one blood every nation of men to dwell on all the face of the earth" (Acts 17:26). I am one with all other humans; it is the doctrine of creation, not star stuff in our veins, that makes all men natural brothers. The doctrine of star stuff makes of me an animal.

## Hope and Change

Words can drive thoughts and imaginations into deep waters, but you should never wade into water where you cannot see the bottom. Evolution, a slippery word charged with hope and change, is a bottomless pit. Christians face charges of stupidity or cupidity for believing God's word or for thinking it incompatible with evolution. To be sure that the accusation is fair one should really define what the word means, lest rebuke is the product of a straw man.

In modern popular use evolution equals change and change equals evolution. The hypothesis of chemical evolution assumes that life sprang from random chemical reactions of nonliving compounds, and the hypothesis of biological evolution proposes that new and different life forms, with increased complexity and genetic informational content relative to the parent, arise due to random genetic mutations. One would hope that scientific discussions of evolution would use scientifically careful language but scientists may unfortunately often use scientific terminology in a popular sense.

A danger of using the word *evolution* in the popular sense is that someone who accepts the idea of change as being a reality of life may soon equate the idea of mere change with evolution, namely biological evolution. Even among an educated populace, the continuous repetition of a falsehood will ensure that, in time, it becomes folklore that most people believe. Does the weather evolve? Does our knowledge evolve? Did my hair color evolve? Do animals evolve? This way of defining the subject makes it possible to characterize any example of change as "evolution," and any opposition to evolution as downright stupidity.

An abstract from a paper presented at a scientific conference sketches the problem: "Evolutionary change is a powerful framework for studying our world and our place therein. It is a recurring theme in every realm of science: over time, the universe, the planet Earth, life, and human technologies all change, albeit on vastly different scales. Evolution offers scientific explanations for the age-old question, 'Where did we come from?' In addition, historical perspectives of science show how our understanding has evolved over time."[309] Simply replace the word *evolution* with *change* and you will see through the silliness. In logic it is called the fallacy of equivocation.

Clearly those who know the code words can brand anyone questioning evolution as an idiot. Do you deny that human knowledge evolves? Do you deny that Earth evolves? It is plain for all to see that all these involve change; they evolve. Use the word as "change" or "develop" and even Christians can easily agree. Well then, can we not agree that life also evolves? At this point one may become cagey and begin to insist on a proper definition of words. If one were to rely on logic and rational argumentation alone, then the battle would end there, for evolution is patently not simply change.

The previous quote illustrates how the word *evolution* can become a propaganda tool when used in a popular sense and a semi-scientific sense with no care to distinguish how these differing ideas overlap. One may happily admit there have been historical changes in diet among different cultures and even in physical characteristics such as height, which diet and breeding may influence. But if you suddenly dig in your heels and need clarification and definition of words when required to admit that your ancestors are pond scum, the reaction will be incredulity, or worse, hostility. By sleight of word and clever semantic gymnastics one moves quickly from agreement that the weather indeed experiences change and human beings experience change, to my ancestor was a prawn, or something equally ridiculous.

## Man and Beast

In the history of the creation one finds ample evidence of the chasm between man and the rest of the cosmos. At God's command living creatures sprang to life fully formed from the waters and the land; with man it was different. Everything about Adam's creation, as reformer John Calvin wrote, distinguishes man by some "mark of excellence from brute animals: for these arose out of the earth in a moment; but the curious dignity of man is shown in this, that he was gradually formed. For why did not God command him immediately to spring alive out of the earth, unless that, by a special privilege, he might outshine all the creatures which the earth produced?"[310] God made Adam from dust of the Earth, but not in an instant. He did not suddenly spring up whole formed but the Creator lovingly erected him from the earth. And there on the earth the material body lay at rest, still waiting for life through unity with the God-breathed part. Finally, when God made Adam a living soul it was plain from the record that he was both a spirit and a material body. He differed, therefore, from the animals and other living creatures, and certainly from the inanimate stars, in that he had a soul, but they were restricted to the material.

Another curious distinction is the order of his creation. Adam was made last. God made nothing new after him. After God had made all else it was as if there was a pause in the divine counsels, and then something remarkable, like the calm before the powerful hurricane, or the deep breath before a plunge. Genesis 1:26 lifts the veil from the mystery of our creation. Evolution urges awe and reverence because our origin remains mysteriously covered by the inscrutable dust of vast ages, so the story goes. The Bible urges awe and reverence through revealing what would otherwise remain hidden forever: "Then God said, 'Let Us make man in Our image, according to Our likeness.'"

What is remarkable in this statement is that God reveals so clearly the plurality of His own personality, whereas this is not obvious with the creation of other creatures. Furthermore, stars and such, the Earth itself, He created *ex nihilo*. But the living creatures of our planet He made from preexisting stuff. In Genesis 1:11 we read: "Then God said, 'Let the earth bring forth ....'" In verse 20: "Then God said, 'Let the waters abound ....'" And in verse 24, the same again: "Then God said, 'Let the earth bring forth ....'" Inanimate stuff He made from nothing but all the living things of pre-existing materials. Even so with man. But with man, as described in verse 26, He signals an enormous difference: "Then God said, 'Let Us make man in Our image ....'"

Our means of creation is a singular distinction from all the other creatures. Not the water, nor the Earth, but God the Father, Son, and Holy Spirit engages directly and explicitly with man. Of course, God places value on all the things He has made, but the wording is so much more personal when He refers to mankind and suggests, according to theologian Gordon Clark, "a more direct relationship between God and man than between God and the animals."[311] The history and work of redemption places the spotlight on the value of man; it is for the salvation of mankind that Christ came to live and die as a man (Romans 5:7), and God counts man alone as the apple of His eye.

## Superior and Inferior

The Green Dragon circulates the idea that man dies as the beast dies (Ecclesiastes 3:19). It is true that our bodies, like those of the beasts, suffer corruption and fall down to become dung of the earth. But how false and wicked the idea that man is a beast when he is, as Pascal said, "neither angel nor beast."[312] Environmentalists who seek to reduce humanity to this are puffing themselves up to look like strutting Galapagos iguanas.

When we neglect or minimize the fall of man, we fail to acknowledge that "the creation was subjected to futility" (Romans 8:20) and pretend all is well when in reality "the whole creation groans and labors with birth pangs together until now" (v. 22). In light of Biblical data the argument for death as something positive is about as pathetic as the argument made by Monty Python's Black Knight.[313] Though King Arthur hacks off several of his limbs the Black Knight remains optimistic of his chances on the field: "It's merely a flesh wound," says he. In a similar fashion the modern environmentalist movement, once again, gets everything exactly backwards. Only it is even worse than this; environmentalists imagine that life and death, death and life, are natural and inexorable, even good.

Darwinian evolution requires death as a positive force for hope and change, but Scripture states that death is an upheaval against the original order. Death is unnatural and ugly, and Jesus knew it better than we do, better than we can, for He was there in the beginning when sin entered this world. More than this, He came to die (John 12:27) to pay the wages of sin, for "the wages of sin is death" (Romans 6:23). And so we can understand why "Jesus wept," as says the shortest verse in the Bible, when He beheld the grave of sin and this world holding the corpse of Lazarus (John 11:35). Adam's disobedience and rebellion, not evolutionary fairytales, brought calamity to the world. Thus the primary direction of influence is not the Earth on man, but man on the Earth, and this is so because our relationship to the Earth is mainly religious. God first cursed Satan for his role in the fall, then Adam and Eve, and all their posterity, and then the entire cosmos, including all its living creatures.

## Guilt Trip

So Adam sinned. Wracked by feelings he had never known before, a guilty conscience and fear drove him to flee from before the face of God: "So he said, 'I heard Your voice in the garden, and I was afraid because I was naked; and I hid myself.' And He said, 'Who told you that you were naked? Have you eaten from the tree of which I commanded you that you should not eat?'" (Genesis 3:10, 11). Oh Adam, what have you done? He had no idea of the horrors that he had been warned of, for though he knew death was not good, yet he had no example of it to excite his imagination to dread. A single prayer would have caused Satan to fly, or a single remembrance of his heavenly Father's kindness and His instructions to obey.

God did not curse the watching holy angels, as they were chiefly observers in that cosmic theater, an audience with their own story, and not active participants in the immediate tragedy of man. Though Satan also sinned and God broke the serpentine tool, there was no general curse on the Earth for his sin; Satan bears his own guilt but passes none on to any other creature. Then why did God curse the Earth? For it did not cause Adam to sin. Neither did any of the other beasts, yet He cursed all, not only the serpent. It was because of man that the cosmos was cursed, and this displays ever so clearly his relation to the Earth and the rest of the created order. Moral responsibility accrues to man alone. The cosmos, Earth in particular, is the stage on which man plays in the drama of creation, fall, and redemption. Even though it was but the stage on which man acted out his disobedience, it too must become marred. Because it is the stage it must be spoiled.

If the cosmos had some voluntary consciousness, if the beasts did, then their guilt would be real. For why did they not rise and smite the demon-possessed serpent, or sound the alarm? They were passive because this is their nature. And so no one blames the Earth, for it is not a moral being to have a guilty conscience, yet it suffers a punishment. Neither do we blame cows, goats, or sheep. But they must still die.

Man's exalted position compels the Earth to share in the consequences of our first parents' actions. But neither to the Earth, nor to the beasts that died because of the sin of Adam, is there any attribution of guilt. This is no consolation, nor does it make their death any less tragic or Earth's curse more comfortable. But at least there is no guilt, a burden God lays entirely upon the man and his wife. For the beast destroyed for goring a man (Exodus 21:28), its punishment ends at death. Its punishment serves to draw out the preciousness of human life, as vastly above the value of the beast. For the man that has murdered, death is just the end of the beginning of his punishment. For man there is the gloom of a guilty conscience and eternal punishment, and so we are distinguished again, sadly this time, from the beasts.

Scripture proclaims Adam and Eve sinners because they were, unlike the other creatures, moral agents. Their sin was the transgression and lack of conformity to God's law. Eve not only broke the law by eating the forbidden fruit, she positively encouraged her husband to do likewise. And he, loving her more than God, willingly and willfully disobeyed his Creator. Remembrance of this tragedy is surely one reason Jesus warned his contemporaries, just several thousand years after Adam, "If anyone comes to Me and does not hate his father and mother, wife and children, brothers and sisters, yes, and his own life also, he cannot be My disciple" (Luke 14:26). A radical, all-consuming passion for God is the design for the lives of humans. Anything less is not worthy of God. Adam was not worthy, and God cursed him and his posterity to a life of misery and suffering here on Earth followed by death. "[T]hrough one man's offense judgment came to all men, resulting in condemnation," and "by the one man's offense many died" (Romans 5:18, 15).

Yet, praise God, in judgment He remembered mercy, for He is a God who delights in mercy (Micah 7:18). This is where the clothing comes in. Humans do not wear clothing because of anything to do with body lice, the current evolutionary just-so story. We wear clothing chiefly because of sin and guilt, for God's righteous judgment fell also on us, "even over those who had not sinned" (Romans 5:14). As the head and representative of the human race, Adam fatally dragged us with him in that first sin. He tried to cover himself with fig leaves. He no longer wanted the presence of God, but fled from before Him (Genesis 3:7–8). God wrote (3:21), "Also for Adam and his wife the Lord God made tunics of skin, and clothed them."

## Death to Life

How Adam must have mourned his disobedience and shrunk in revulsion from its effects! First, God no longer took pleasure in Adam; this alone was enough to render him inconsolable. We do not know whether God brought a creature and slaughtered it before a horrified and sickened Adam and Eve. But we do know that God killed a creature (at least one) of his own making and Adam and his wife had to wear the rotting skin torn from its bloody flesh. The opening death recorded in the history of the cosmos is that of an innocent beast. By this sign they learned that their nakedness, a strong awareness for their newly earned depravity, was to be covered at the expense of life.

The heavens, the Earth, and all creatures great and small received a brand, as it were, with the mark of Adam's condemnation. God wanted Adam and his posterity to ponder this point carefully and so spent considerable time explaining how the entire cosmos would become marred as a result. God did not explain to the beasts, but to man alone. This is plain from God's first approach to Adam while he was hiding (Genesis 3). God called to him. Then the Lord interrogated Adam. When Adam blamed his wife, God questioned her. But God did not question the serpent, for, as Calvin wrote, "in the animal itself there was no sense of sin, and because, to the devil he would hold out no hope of pardon."[314]

Adam received visible tokens to contemplate the wrath of God, in the spoken curse, in the guilt and ruined relationships that he experienced, in the change in his fundamental nature through the loss of his original righteousness and holiness, in the visible marring of the cosmos, and in the death of the animal whose skin he wore. Lest he should despair God gave him a precious promise to ponder, reminding him that though Satan be a fearful enemy yet the seed of the woman would eventually crush his head (Genesis 3:15). All of this made him look ahead with hope to the day of fulfillment of God's promise. This is a great difference between man and beast, for animals do not reflect on the distant future, or have present enjoyments diminished by the intellectual prospect of their impending death. Goats do not fret over global warming or worry that their kids have a purpose driven life. Unlike the beasts, who only feel the effects of Adams depravity, human beings are also able to meditate on these things and engage "the will in bringing its ideas into contemplation, and ranging and comparing of them in reflection and abstraction."[315]

The Green Dragon holds that appetites drive man as the brute beasts, but God says that man is more than this. Adam had free will but with his fall we became different in that we bear an original hereditary moral corruption. Yes, it is true that man *qua* man is now much less than he was before Adam sinned. In some ways our natural state is now much like the beast and much like Satan, the one being some subtle animal and the other the devil incarnate.

Man has ego, unlike Dawkins's ape. The ape takes no offence whether its master is an Oxford don like Dawkins, or the garbage man. It matters not to him that his master is rich, poor, short, tall, thin, or fat. He is content to serve a master who only feeds his flesh. Modern brain scientists reduce all the difference between man and beast to mere chemicals and synapses and therefore to one of degree, not of kind. But the reason the beasts never speak of eternity and we do is not found in their organs, but in our spirits. As Robert Dabney wrote, "Mind is not a habit; it is the spiritual thing on which habits form."[316] Our fallen nature yet holds knowledge of spiritual truth and our sentience of morality distinguishes us from lower animals. This is why fallen humans still order their behavior consciously, though often wickedly, rather than by mere animal instinct.

## Conclusion

Ancient pagans believed there were spirits on all sides, spirits of nature and place. The ancient Athenians also reverenced beasts or ascribed spiritual qualities to nature, but they still remembered that they were themselves the children of God. Modern environmentalism casts back in time to what environmentalists imagine as the deepest deep. They cast God away, and reel in the beast as parent. They commune with the tree or the untamed blue sky because they do not understand that man's prime relationship is fundamentally vertical, upward to a personal God. They see like cause with the beast as a fellow creature, but fail to see that spirit elevates man above the beasts. God created man with the capacity for communion with Himself. Pascal wrote: "I condemn equally those who choose to praise man, those who condemn him, and those who live for themselves."[317] He is no mere animal, nor an angel.

Unreasoning animals are totally immersed in the moment. Aldo Leopold's great Gray Wolf high on the ridge is a case in point. Before the green fire in her eyes burned out she had been attuned to the scents carried by the breeze, the rich wildness of the world and her part in it. No doubt, she did not care to meditate on the meaning of it all. But she was alive to the immediacy of each moment, every nerve tingling to the nervous energy coursing around and through her. This is what environmentalists of all stripes urge for humanity: recognition at an irrational animal level that all creatures share a collective reality or consciousness with all matter and energy. It is a sense of feeling beyond reason. But it seems to me that environmentalists place on animals a burden they are incapable of carrying. They are not reduced humans, they are beasts.

The narcissistic impulses of modern civilization, enamored with Hollywood-style religion-lite, find this religion of self momentarily satisfying. Unreasoning mystical communion with an unrevealed and unknowable God—call it Buddha, call it baby Jesus, call it Gaia, or just plain old Mother Nature—demands, as pundit Spengler wrote, "the sort of star-and navel-gazing that brings the communicant right back to good old number one."[318] In the end, the new religion that Greens embrace boils down to the temptation of the garden of Eden, to be number one, to see God as an idealized image of ourselves.

In summary, the material creation is not divine or eternal, neither of intrinsic value nor in unity with mankind. If unbelievers cannot lift nature by lowering man, then perhaps they may hitch it to man so that they are exalted as a team. But Christ and the whole Bible teach that man is a creature of God whose ancestry is not animal, but divine, and that God made the Earth for man. Those who have so vigorously argued that man is a naked ape, an animal borne on cosmic tides, are toying with a deadly creature. They are children playing with serpents. Or is it the serpent toying with children? C. S. Lewis voiced the horror of this in his *Prince Caspian*, in which the little girl Lucy asks, "Wouldn't it be dreadful if some day, in our own world, at home, men started going wild inside, like the animals here, and still looked like men, so that you'd never know which were which?"[319]

# CHAPTER FIVE

## Dominion:
## How Humans' Filling and Ruling the Earth Can Release it from Bondage

*When you have eaten and are full, then you shall bless the LORD your God for the good land which He has given you. Beware that you do not forget . . . lest . . . your heart is lifted up, and . . . you say in your heart, "My power and the might of my hand have gained me this wealth." ... Then it shall be, if you by any means forget the LORD your God, and follow other gods, and serve them and worship them, I testify against you this day that you shall surely perish.*

Deuteronomy 8:10–19

## The Source of Evil

Imagination quivers at the Hindu-inspired horror described by Stanford University butterfly expert Paul Ehrlich: "The streets seemed alive with people. People eating, people washing, people sleeping. People visiting, arguing, and screaming. People thrusting their hands through the taxi window, begging. People defecating and urinating. People clinging to buses. People herding animals. People, people, people, people."[320] Ehrlich oozes insufferable condescension for the grubby city of Delhi, India, especially for its swarming people.

One wonders at his impressions of Monaco, that European playground, whose population density, at nearly 44,000 people per square mile, is about fifty percent more than Delhi. Though he has expressed a general animus towards overpopulation, one suspects he likes Monaco much more than Delhi. The idea that people should dominate the landscape oppresses Ehrlich, and so he rails against it. In this chapter we delve into his discomfort and test and contrast his ideas against the Christian doctrine of dominion.

It is ages since Christianity penetrated that world of European barbarians, transforming and renovating their culture and institutions through the sanctifying influence of the gospel. Though still paying lip service to the Christian meme,[321] by now Western culture has largely regressed to pagan narcissism, the degraded ethic of me, Me, ME. Western pagans nevertheless still enjoy and covet the benefits produced by application of Christian values, though current trends suggest not for much longer unless there is genuine spiritual revival.

Values and truths that once uplifted old Europe and largely transformed the continent from paganism can transform Delhi to be healthy, happy, and holy. Even critics of Western Civilization like Ehrlich involuntarily admit its excellence. Though tempting, it would be false to credit Ehrlich's attitude to cultural chauvinism or unfamiliarity with the brown other. Delhi is a frequently unpleasant place in which to be foreign, and its bustle and hustle can be unsettling. But the population density is not much different from a typical American shopping mall—which Ehrlich also hates. Despite Ehrlich's distaste at the putrid press, modern Delhi is a vast improvement from what it was before Christian influence. Yet still today the life of a child is especially cheap there;

pagan cultures devalue life, and some even codify the sacrifice of human life as a moral or religious virtue.

The Hindu caste system relegates people born into the Dalit, or Untouchable, caste to the lowest form of humanity, and many hold them as subhuman. According to caste rules if a Dalit's shadow falls on an upper caste person, that person becomes polluted. In contemporary India their lives are almost worthless.

Hindu ethics is not an ethnic aberration. Such neglect for life litters the graveyard of human history. Rome turned killing into circus sport. Ancient Canaanites sacrificed their children to their idols. Mayans, Aztecs, and Incas turned human sacrifice into bloody religio-political theater.[322] Degenerate twenty-first century Western Civilization enshrines the murder of unborn babies by planned abortion as a high form of self-expression. Yesterday's expressions of Satan's tender mercies trend on the same themes in view today. Paganism, whether ancient or modern, does not raise but rather degrades culture. Paganism kills.

Christianity holds the antithesis, and the moral high ground. All human life is precious, for each one of us is God's image, and infinitely valuable in His sight. We dare not quench human life on the feeble authority of hunger, societal whims, or traditions, but only by the express command of God. Therefore Christian laws value human life, support activities that cherish and celebrate life, and naturally oppose murder or suicide. It was primarily Christians who managed successfully to stigmatize the ponderous pagan custom of sati.[323] Sati, a woman's immolation on her husband's funeral pyre, once conspicuous of upper caste Hindus, is now nearly absent from the country. Much is the pity that Christian influence has not penetrated further into the warp and woof of that society. Had the glorious freedom in Christ captured that spiritually enslaved culture India would now be a far more momentous power, exporting Christian missionaries to Europe and, yes, especially to the United States of America.

Paul Ehrlich has long recognized the role of Christianity in transforming cultures, breaking the ice, shattering the still and unencumbered space he so covets. Yet Ehrlich speaks with contempt and revulsion not only of Eastern but also of Western Civilization. Perhaps he is simply a misanthrope; in which case his claims to love nature are not because he loves it but because he wishes to cover up his hatred of man by fine-sounding platitudes. One can hardly resist a bittersweet laugh at Ehrlich's expense; he is an environmentalist Dr. Strangelove whose hatred cuts him off from God's good love, but who cannot help lusting after natural exemplars of divine glory.

It is difficult to shake the impression that humans are for him a constant irritation, an unscratchable itch, an ingrained filth snared under fingernails. Christianity is, for Ehrlich, the insulting inertia that stares at him with heavy, limp-lidded gaze, covering and protecting the dark obscenity that is mankind. Scratch, scratch compulsively as one may, it remains.

Ehrlich is famous for obsessing about mucky humans; this focus is largely responsible for his elevation to iconic status in the Green movement. Al Gore counts him as a Green giant, perhaps even giving kudos by naming his *Earth in the Balance* for an Ehrlich quote. Anyone that has read and thought about Gore's book finds the similarities shared with Ehrlich's work striking. Ehrlich helped animate and invigorate the Green movement in the United States with his 1968 book *The Population Bomb*, which rode the back of the hippie movement. Wrote he, "Careful plans must be laid for getting the Earth back in balance."[324] For billions of years our Earth, fragile as Humpty-Dumpty, spun with eccentric poise. But today the planet is off its ecological hinges, so goes the Green narrative, because Johnny-come-lately humans exercise dominion over it.

His prescribed solutions major on the theme that humans are going to have to die; not millions, but thousands of millions will need to pay with their lives before the Earth will be right again. In the end, for environmentalists, the natural world is a world without man. Dominion is thus a word that, to the radical environmentalist, is fraught with moral hazard. It usually means the power or right of controlling and governing. Or in legal terms, it implies both title and possession. Such lordship, when earnestly believed or exercised, can be threatening or distasteful for those who see man as an insignificant pinprick on the inflated canvas of time, but one whose sharpness threatens to burst nature's heart. In *Population Bomb* Ehrlich wrote, "Our entire system of orienting to nature must undergo a revolution. And that revolution is going to be extremely difficult to pull off, since the attitudes of Western culture toward nature are deeply rooted in Judeo-Christian tradition. Unlike people in many other cultures, we see man's basic role as that of dominating nature, rather than as living in harmony with it."[325]

Whether Ehrlich's analysis of pagan and Christian culture is accurate, it is clear that the idea of God's emphasis on the special creation of man is unpalatable. Especially unappreciated is God's command that Adam and Eve multiply and exercise dominion, which Ehrlich falsely equates with domination, over other creatures. Here is what the Master of the Universe says (Genesis 1:26–28): "Then God said, 'Let Us make man in Our image, according to Our likeness; let them have dominion over the fish of the sea, over the birds of the air, and over the cattle, over all the earth and over every creeping thing that creeps on the earth.' So God created man in His own image; in the image of God He created him; male and female He created them. Then God blessed them, and God said to them, 'Be fruitful and multiply; fill the earth and subdue it; have dominion over the fish of the sea, over the birds of the air, and over every living thing that moves on the earth.'"

## Man on Man

Ehrlich bristles at Biblical language and argues that ideas of human entitlement and supremacy over nature are invalid and harmful. To keep Humpty-Dumpty from having a breakdown he recommends solutions that bear grave and unappetizing likeness to those tired and failed utopian ideas of past revolutions. At the very least, he prescribes a reevaluation of Christianity. Here is what he considers the Christian position: "God designed and started the whole business for our benefit. He made a world for us to dominate and exploit. Our European ancestors had long since developed the 'proper' attitudes when the opportunity to exploit the New World appeared. Both science and technology clearly have their historical roots in natural theology and the Christian dogma of man's rightful mastery over nature."[326]

Ehrlich's references to the "New World," in so many words, direct one to equate human superiority over the natural world, especially the attitude that our mastery is proper and inevitable, with imperialism. Christianity, he believes, provides the impetus for this ecological "imperialism." Ehrlich's arguments drew on earlier works by Henry

Fairfield Osborn and William Vogt and faithfully track those of UCLA historian Lynn White in his influential 1967 *Science* article "The Historical Roots of our Ecologic Crisis." Said White, "Formerly man had been part of nature; now he was the exploiter of nature."[327]

Thinkers such as White and Ehrlich understand that men are prone to wickedness, for instance our fatal attractions to war and real imperialism. But they underestimate the role of a bad heart. Our alleged wicked attitude towards nature they consider a byproduct of Christian culture that once dominated Western thought, and today remains in residue. Take, for instance, outrages against American Indians in the "New World". Ehrlich claims genocidal impulses are exacerbated by ideas[328] of dominion. It is convenient to blame outrages on the Christian principle of dominion. That is but a straw man as is obvious when one considers how Indian on Indian outrages, common before and after the arrival of old world settlers, receives rare mention. It is now commonplace to imagine that before the pilgrims came life in North America was somewhat idyllic, and the Indians did not dominate anything since they were not in thrall to Christian ways of thought. They lived in perfect green harmony with the Earth, the trees, and adorable furry friends.

Facts are hard teachers. In Alberta, Canada a cliff as high as 30 feet and 300 feet long is the place where Blackfoot Indians drove herds of bison over the precipice. Those animals that did not die outright suffered broken legs, backs, and other horrific mutilations before the final orgy of slaughter *en masse*. The crag has a poetic name, translated into English as *Head-Smashed-In Buffalo Jump*, for the foolish boy who lay close to the cliff face to watch the hordes of bison stampede over the edge. So great was the carnage that as the bison piled up the boy became pressed against the cliff, where his soft head soon was smashed in by the growing scrum of bison carcasses. This is surely not the nature harmony culture a good Christian would emulate.

## One Blood

To blame Christianity for distorting the Biblical doctrine of dominion is to blame mountains for climbing accidents, or gluttony on food. Ehrlich pretends to be outraged at abuses supposedly mothered by Christianity, but perhaps he is really just outraged by Christianity. In point of fact, Christianity acted as a restraint on the lusts of men with guns and steel and ameliorated the rapaciousness of their advance against the Indians, who were themselves arguably superior in savageness[329] but unquestionably inferior in killing technology.

George Washington, first President of the United States (1789–1797) and a man with a strong Christian sense of morality, tried in vain to stop various shameful and illegal land grabs and other provocations against Indians. He believed the Christian maxim: love thy neighbor. In correspondence to his Secretary of State, Timothy Pickering, he lamented in 1796 that "scarcely anything short of a Chinese wall, or a line of troops"[330] would hold back those intent on theft of Indian land.

God's Word teaches the Christian view that private property rights exist, and thus deserve respect—"Thou shalt not steal." God's law thus requires the existence of private possessions. Yet there will always be those who would twist any good, even the Scriptures, to their destruction (2 Peter 3:16). Some people said the American Indians were not humans but savage beasts. Thus stealing their land was not theft, but the proper exercise of dominion over animals. In this way curiously creative, but fundamentally dishonest, people are thus able to call good bad and imagine that bad is good. How convenient. And how akin to the pagan Indian view of low-caste Dalits.

Men like David Brainerd (b. 1718) and Jonathan Edwards (b. 1703) devoted themselves to serving the Indians. They did so precisely because of their Christian view of dominion, in particular their role in exercising the dominion of Jesus Christ to disciple all nations (Matthew 28:19). While upholding the innate dignity of all humans, they had no scruples about calling the natives heathens, pagans, savages, and barbarians. Unlike humanist love, which pretends that all cultures are equal, that diversity is desirable in itself, Christian missionary love wants what is genuinely good for others, namely God's grace in their conversion.

They recognized too that all men without Christ are monsters whether dandified in Western clothing or naked and howling at the moon. Mao, Stalin, Hitler, Che, Pocahontas, you, and I are as alike as peas in a pod, from the standpoint of the perfect righteousness of God. This is not to say that my baby daughter is as wicked as Mao the monster. Without a doubt she does not have the blood of tens of millions on her hands.

She is at least innocent of that. But she is no better in her infancy than was Israel's King David in his, who confessed himself to have been sinful from conception (Psalm 51:5), and therefore capable of murder. All of us, from the least to the greatest, face doom unless God saves sinners, "for all have sinned, and come short of the glory of God" (Romans 3:23).

This means there is no such being as Rousseau's noble savage, whether on Wall Street or anywhere else. Brainerd and Edwards believed that God had made of man a unity. Christians teach that God "has made from one blood every nation of men to dwell on all the face of the earth" (Acts 17:26). We were all once heathens, dead in sins and trespasses. We are all absolutely reliant on the mercy of God, and all stand condemned by His infinite justice, yet even so all are precious for God makes all (unlike the animals) in His image.

Brainerd spoke of one Indian convert who "seemed to be filled with love to Christ: at the same time she behaved humbly and tenderly, and appeared afraid of nothing so much as grieving him [Christ] whom her soul loved."[331] Secular society considers such humble submission before an invisible—and therefore nonexistent—God an affront and a strong indictment of the Christian way. They see only a noble savage reduced to groveling before shadows because of feelings of guilt. However, what we see expressed by this woman is not weakness, but a great strength of Christianity. She recognized her total vileness and need for Christ. We are all equally culpable before God's justice and, like it or not, will all one day bow the knee before Him. The Green Dragon eschews bowing and instead encourages the return to our self-reliant pagan, heathen, barbarian, savage roots and, unlike this woman, bears no love for Christ. How is it with you?

Brainerd literally spent himself for the Indians, exhausting his frail flesh and passing into glory at age 29. Edwards preached for five years to the Housatonic Indians and defended them against thieves and robbers, those so-called Christians who wanted to relieve them of wealth and dignity. He strove furiously against colonists who used their political influence to enlarge their private fortunes at the expense of the Indians. He did not do so because of *faux* multicultural sensibilities, or exquisite feelings of white man's burden, but precisely because his God had commanded him to love his neighbor, the image of God.

Jesus Christ changes hearts, ennobles humanity, and so restores to us the knowledge of our true relationship to one another and to God.

## Eco Imperialism

Environmentalists equate dominion with domination, a word chock-full of negative connotations. Then they charge this straw man and use it as an excuse to rail against the true religion. Once crafted the straw man makes it possible to accuse selfless missionaries, like Brainerd and Edwards, of being imperialists for shaking the authenticity of the natives. For instance by clothing their nakedness, or teaching them the glorious gospel. Presumably, then, Jesus too was an imperialist.

Protestant author Wendell Berry, a thoughtful environmentalist, rejects the view that every culture, like every knee, must kneel before the lordship of God. He asks that we consider that "the complicity of Christian priests, preachers, and missionaries in the cultural destruction and the economic exploitation of the primary peoples of the Western Hemisphere as well as of traditional cultures around the world, is notorious."[332] This would be a valid point were he arguing simply against the idea that it is peculiarly Christian to abstain from moccasins in favor of high heels. It is true that some Christian missionaries have vainly taught their traditions as if they were the oracles of God. It is also true that many missionary societies that claimed to be Christian, but were not, abused the language of Christianity as a means to exploit and enslave natives. They wanted to extend man's kingdom rather than the Kingdom of God. This is not so different from the contemporary behavior of many so-called ecotheologians in the emptying churches of the West. But Berry argues against more than mere fashion imperialism and real abuses; he argues that it is somehow sinister to replace native cultures by Christian culture.

The difficulty, of course, lies in that it is not always easy to define one part of culture as Christian and another as anti-Christian. It is not necessarily pagan to eat meat, or is it? Or to eat only organic produce, or consult naturopaths? Our method of answering such important questions can lead to absolutes and clarity, or relatives and cacophony. Were we to rely on tradition or church authority an endless list of changing obligations would soon burden our view. How then could one believe one's tradition is not necessarily inferior, or superior, to another's? One man's pope speaks *ex cathedra* and another's guru guide speaks with a magical puff of smoke. If popes and gurus all pointed to the same reality one could at least sympathize with Berry's discomfit at the unbending certainty of those who claim Jesus alone is "the way, the truth, and the life" (John 14:6). If all cultures are fundamentally equivalent, then it would not matter much in practical terms who was "right." Truth is the first sacrifice on this altar.

## Truth in Christ

Just because it is sometimes difficult to come to Biblical answers to specific issues does not mean it is impossible, or that clear principles do not exist. Life is complex. But even so we can still find answers to difficult questions. That is why God provided His Word, to provide light on a dark path (Psalm 119:105) as a guide in leading a holy life. The judgments and guidelines God provides via His Word, and application of these words by His Holy Spirit, warn His servants how they should frame their life (Psalm 39:4). As Jesus said so succinctly (John 17:17), "Sanctify them by Your truth. Your word is truth." Culture, like the souls of Christ's disciples, must undergo constant renovation away from sin and towards righteousness. Our life should be one of constant stretching for a heavenly goal, a goal that extends Christian dominion from the personal, individual level, to the realms of family, church, state, culture and nature.

If God is alive, as we assert, Christian culture should replace native cultures, and in the end it will, from sea to shining sea (Psalm 72:8). That is what Christ was driving at when He prayed, "Thy kingdom come." Christ's eternal dominion will soon replace all other kingdoms. Even now they shrivel steadily. Gospel power will renovate every culture. Except for rejection of pagan gods and ideals, how else will *every* knee bow and *every* tongue confess that Jesus Christ is Lord, as the Scripture so plainly confesses must occur (Philippians 2:10–11)? When pagan gods, which are no less idols than the Philistine Dagon, bow, their unity shatters (1 Samuel 5:2–4).

God calls heathen culture an abomination. It is not that they eat the wrong food or wear the wrong shoes, though these items are not necessarily unimportant (Acts 15:29). It is that they bear no love for Christ and so, says the Scripture, they "love death" (Proverbs 8:36).

And what is a love of death but a self-hatred, thus a hatred of man and all that he does. Jesus said the disciples must teach new followers "to observe all things that I have commanded you" (Matthew 28:20) and to reject their former paganism, "as obedient children, not conforming yourselves to the former lusts, as in your ignorance" (1 Peter 1:14). One cannot honestly escape this teaching by caveats and reference to the sins of others. Berry's argument is contrary to Christ's, therefore false, and further rests on a logical fallacy. One may as well argue that paramedics and ambulances are bad because of shyster ambulance chasers.

Berry, like so many professing Christian environmentalists, rejects the exercise of godly dominion. To him sinful motives and behaviors irremediably taint man's control over nature, and he thus views dominion as illegitimate. The former foul taint is inescapable, the latter conclusion illogical. Ehrlich takes it a little further. Human dominion, he believes, is part and parcel of the Christian religion. It repulses him. However much his views caricature true Biblical teaching, Ehrlich's frank reactions display evidence that he has at least grasped the fundamental antagonism between Christianity and environmental fundamentalism. Here he is much more honest than his intellectual posterity, particularly those professing Christianity who are alive to the possibilities of transforming and co-opting Christianity, suitably redefined, to energize the environmentalist cause. They abuse the Scriptures to answer to the passions and prejudices of the Green Dragon.

## Self Image

In *The Rights of Nature*,[333] environmental historian Roderick Nash has ably cataloged how, from genesis to the present, the radical Green movement is shot through with anti-human thinking. Nash explains how the Green Dragon wishes our ethics to evolve to engage with the expanding environmental consciousness. With Green philosophy rising to the point of social and political policy we must remake our actions and thoughts.

Lynn White explained precisely how to achieve this goal. He wrote, "What people do about their ecology depends on what they think about themselves in relation to things around them."[334] White cites with approval the example of Francis of Assisi who, he claims, "tried to depose man from his monarchy over creation and set up a democracy of all God's creatures." White says that Francis was clearly heretical by the standards of the day. But let heresy be damned—White calls Francis the revolutionary who provided the foundation for a reinterpretation of the basic Christian principles of human dominion; Francis called the animals, even the inanimate world, his brothers and sisters, and he was not merely waxing poetical.

"As things stand," declared spiritualist David Spangler, "before we can manipulate or exploit something or someone, we must first devalue them, making them lower than ourselves. That which is sacred cannot be devalued, and by naming the Earth and all upon it as sacred, we seek to protect it and ourselves from ourselves."[335] In this case, at the least, the religious connotation words and Gaia terminology, even if used only as metaphors, make it difficult to view the Earth as dead matter of which man is the bumbling governor.

## Holy Things

Baptist writer Wendell Berry insists that Christians have lost a sense of the holiness found in all of creation. One reason, says Berry, is that Christians have falsely associated the built church as the only holy place. Berry holds that all is holy because God is present in all places, a view Al Gore enthusiastically endorses. It sounds appealing until one realizes that all is *not* holy, as the bare feet of Moses knew so well. Neither is hell holy, though God be present there too. He is omnipresent, after all, and hell is a place, is it not?

Indeed God is omnipresent, but His presence does not automatically imbue every part of the created order with holiness, as if He were a lucky charm to rub up against purringly. The latter is a pantheist position. Because it is consciously and clearly couched in Biblical language, Berry's thought form is all the more dangerous to gullible Christians. He is correct in pointing out the fatal flaw of holding unbiblical positions such as that the brick and mortar at a worship site bear exclusive rights to holiness. Of course, the Biblical position is that God defines holiness. The church meeting house is far from the only holy place. It is not holy at all.

Green philosopher Erazim Kohák, like Berry, insists the idea of sacred is the blunt instrument people use to justify treating the nonhuman world as something without much spiritual importance. By calling Jews pigs and dogs,[336] as does the Islamic Q'uran, for instance, some justify genocide. Kohák argues that when we desacralize animals we are doing something similar—"only animals" he writes, is somehow morally equivalent to "only Jews." This is how the German National Socialists (Nazis) and the Islamic Grand Mufti justified their extermination programs, and how the land jobbers General Washington execrated justified stealing someone else's land. "The idea of human superiority over other animals," says Kohák, silky smooth as a pig's ear, "is clear evidence of racism."[337]

One can hardly resist rolling on the floor with laughter at the soaring idiocy of such a statement. Who would have thought that academic social sciences would become an epicenter for serious comedy? But let us not question Kohák's sincerity; he is not alone in thinking like this. Should society abolish ivory towers because their erection into the powder blue sky is clear evidence of sexism? In such fashionable silliness we see the hollow vapidity of flabby ideas that shock and awe vacant minds with their impressive pseudopregnancy. Man never gets to define the sacred, God does, and no way is the Earth sacred or in any way part of the human race.

## Brother Sun, Sister Moon

Mere say-so does not make the Earth sacred. The argument hinges on a disagreement over what is the proper role of mankind relative to the rest of creation. Nature, radical environmentalists insist, has "intrinsic value." So we should revere the Earth for its own sake, irrespective of any benefit to man, irrespective of what God commands. We must therefore reevaluate our position relative to the beasts. If they are brothers and sisters, sharing common ancestors, then should we not respect them as we do our kin? If we will not revere nature should we not count it an equal, at the very least?

A consequence of stumbling into such thinking is to restrict man from using nature for his own ends. Since nature supposedly personifies value and goodness, human actions that change the environment are immoral. Just as a social adept would not willingly invade another person's space, so we should not crowd into the "space" belonging to any other creature; thus the environmentalist litany.

Christopher Manes, author of *Green Rage*, believes, "Perhaps the most important and lasting legacy of radical environmentalism is that it has raised the possibility that there is another path for humanity besides the ecological imperialism of civilization and the destruction that it brings."[338] If we must live, he would like us to linger lightly on Earth and renounce any megalomaniacal humanist ideas about the crest of evolution, and Christian ideas of mastery and dominion over creation must certainly be discredited. We humans are not special, period.

Here the religious, or faith-based, environmentalists almost willfully tie themselves in knots. For instance, the Presbyterian Church (USA) argues for "the impossibility of separating the human creature from the rest of creation."[339] They are not speaking about the obvious material properties and habits we share with much of the created order. Instead they are reiterating Manes's chief point: stop thinking you are big stuff and realize you are ultimately not more or less valuable than a rat, a pig, a dog, or any other of God's children. Charles Darwin once conjectured that "animals, our fellow brethren in pain, disease, suffering and famine … may partake of our origin in one common ancestor—we may all be melted together."[340] Humans and animals are joined at the hip. Hip theologians surf the Green wave to take Darwin's proposition and turn it into a passion.

The dominant cultural narrative holds that the universe has existed longer than anyone can count, the Earth for close to five billion years, and humanity a mere blink in time. It is not surprising that where this perspective blossoms, like a fast-growing weed, it is common to characterize Christian ideas of dominion as arrogant, even blasphemous. It would be as if Lucifer, a mere created being, would claim to be like God. Says Wendell Berry, "By denying spirit and truth to the nonhuman Creation, latter-day proponents of religion have legitimized a form of blasphemy."[341]

## Hell in a Handbasket

Though constantly advertised as steely eyed realists heroically defending truth and reason, mainstream environmentalists are, as Lord Christopher Monckton put so eloquently, "bed-wetters."[342] There are real environmental problems, but typical environmentalists obsess over hypothetical future catastrophes. And they pooh-pooh all the hard scientific data marking an improved natural environment and life quality for those humans who live in societies that are open, tolerant, and friendly to the free exchange of "knowledge, technology, people, and goods."[343] Rather than finding comfort and hope in God's Word, environmentalists fret that the world is burning up and going to hell.

They ignore or minimize the cool and steady analyses of people like Julian Simon, Bjørn Lomborg, Indur Goklany, and others who have shown that economic growth can lead to a healthier environment. Such is the case because free people that have reached the point when they do not have to spend all their time and energies to fill their bellies and keep warm are able to develop an interest in stopping environmental deterioration that reduces their quality of life. In the environmentalist analysis, all such facts go by the wayside. When we rely on politicians and environmentalist organizations for information it is hard to shake free of the widespread sense of impending doom.

Environmentalists, like aficionados of the horror movie genre, seem to derive a grave pleasure in imaginary scares. As David Goldman has written, "After rejecting revealed religion, modern people seek a sense of exaltation in nature, which is to say that they revered the old natural religion. If you do not believe in God, quipped G. K. Chesterton, you will believe in anything. It is too fearful to ponder one's own mortality, so the Green projects his own presentiment of death onto the natural world. Fear for destruction of the natural world—trees, whales, polar ice-caps, tigers, whatever—substitutes for the death-anxiety of the individual."[344] Terror haunts them deeply into bloodshot nights and nails dug into sweaty palms; humans are, of course, almost always to blame. Instead of awe at the image of God, what fills them is dread and rage. Instead of seeing man as a special creation, they see man as the special problem. For them good intentions do not pave the road to hell, but screaming human babies.

Environmentalists are right about some things. If man is at the peak of creation, and man is naturally wicked, then does it not stand to reason that both men and the rest of creation will suffer damage? There is no denying our sinful nature, or that we sometimes behave terribly to one another and the other creatures. I can think of no person who denies this. Cautions about dominion that begin here are on target. Yet the Biblical doctrine of sin is precisely what most of modern society, along with the Green Dragon, denies.

## Cursed Cosmos

At first it therefore appears odd that people dismayed by the traditional Christian concept of sin map it so effortlessly into the idea of sins against the environment. Lust is not necessarily a sin, nor infanticide, but cutting a tree down may be. Remember the thoughts of environmentalist Næss, whom we encountered in Chapter 4, who explains: "[I]t is against my intuition of unity to say 'I can kill you because I am more valuable' but not against my intuition to say 'I will kill you because I am hungry.' In the latter case, there would be an implicit regret."[345] For Næss there is no sin in murder, only a type of hunger.

Sin may not actually be a factor intrinsic to our nature but an artificial construct that we learn from religious and other social conditioning. There is therefore no ethical connection between our relationship to Earth and the association we, and the Earth, have to Adam. That belief causes environmentalists to idealize nature as an intrinsic good—they refuse to admit that God has cursed the cosmos, that its present natural state is not good. And when professing Christian environmentalists mention the Fall, they fail to take seriously the noetic effects of the curse. Instead they ascribe curses via direct actions of mankind in working the Earth.

Eden was a special place no doubt, but the wilderness is not Eden. Even prior to the entrance of sin, wild nature is not paradise. It is wilderness. Utopians imagine otherwise, but they may as well try to cover a broken windowpane with a sheet of ice. Wendell Berry tries: "It is understandably difficult for modern Americans to think of their dwellings and workplaces as holy, because most of these are, in fact, places of desecration, deeply involved in the ruin of Creation."[346]

My clearing a field is thus imperialism, and plowing it is rape. Planting on it a genetically adapted crop is incest, and using pesticides is abortion. Paving it is genocide. There was no ruin before man, but ruin arrived with the evolutionary emergence of *homo sapiens* a few short years ago. The natural is so idealized and exalted that the idea of man improving the land is, by environmentalists, counted as an oxymoron and abomination. Instead we need to learn to live with it, as a fellow creature of equal intrinsic value. Christ's church can easily fall prey to Green philosophy, not realizing the danger in the tempting fruit. Do not bite till you know whether it is fruit or a stone.

We hear, for instance, that it is better to eat organic because it is natural. And why? Because natural is ostensibly pure, just the way God made it, and unsullied by the "improvements" of man. Never mind that God has cursed the Earth so that it is permanently marred. Natural is not necessarily nice; so-called organic is not necessarily best. Never mind that on a per capita basis modern farming techniques are ecologically far more sensitive than that of ancient Earth Mother cultures. Humans have long used pesticides to prevent damage to their crops. For instance, the Sumerians about 4,500 years ago used elemental sulfur dusting for pest control. By the 15th century, humans applied to crops toxic chemicals such as arsenic, mercury, and lead.[347] We may thank the Lord, for modern agriculture is able to learn from such a toxic past. In Chapter 6 we will explore some of these points further.

## God's Dominion

What is the Biblical teaching on God's dominion? God rules with absolute power, as His written revelation teaches throughout. Consider, for instance, Isaiah 46:9–10: "I am God, and there is no other; I am God, and there is none like me, … My counsel shall stand, and I will do all My pleasure." Or consider again Psalm 115:3: "But our God is in heaven: He does whatever He pleases." His power and authority so exceed that of mere earthly potentates that their authority before Him is as nothing. Therefore Scripture says He "is the blessed and only Potentate, the King of kings and Lord of lords, who alone has immortality, dwelling in unapproachable light, whom no man has seen or can see, to whom be honor and everlasting power" (1 Timothy 6:15–16).

It would be terrible were a mere creature to arrogate to itself godlike authority; many human rulers have, to enduring lamentations. But we remember earthly kingdoms come and go, and kings are just shadows. The kingdom and rule of God is eternal: "Your kingdom is an everlasting kingdom, and Your dominion endures throughout all generations" (Psalm 145:13).

His dominion serves as the supreme reference along which one may steadily train the sight, as along the shaft of a straight arrow, in order to strike accurately the true Biblical understanding of human dominion over nature. Since we are God's image we may hope that a better understanding of God will lead naturally to a better understanding of ourselves. Perhaps the main point to note here is that God, who created the entire cosmos, is Himself without beginning and end or, as Scripture says, "from everlasting to everlasting" (Psalm 90:2). He is in this altogether unlike His creatures, all of which have a beginning, many of which have an end.

We need other creatures, but God is independent and has no need of anything from the creature. Thus in the beginning He created everything of nothing. He made nothing in vain (Psalm 104:24, Isaiah 45:18), and He took pleasure in the creatures that He had made. Indeed, everything created was created for His pleasure (Revelation 4:11). And all these, ourselves included, depend upon His mercy for their moment-by-moment existence. God's mighty Word teaches that it is the second person of the Godhead, Jesus Christ, who is "upholding all things by the word of His power" (Hebrews 1:3). All of creation, from the fragile to the ponderous, should immediately cease to exist if God did not continually exercise His will to uphold and sustain. This puts the lie to pagan materialism, which holds that matter is the whole enchilada when, in fact, matter comes from spirit. Or, as Jesus said, "It is the Spirit who gives life; the flesh profits nothing" (John 6:63).

God certainly has power to command and govern His creatures. Not only does God have the power to carry out His will, but He also has the right to govern. He founds this authority upon His decision to create and carry everything: "God, who made the world and everything in it … is Lord of heaven and earth" (Acts 17:24). In other words, He owns us.

Because He made all things, He is properly titled Lord, and not just of some but Lord of all (Acts 10:36). Things that depend wholly upon God for existence, and this means all things, are naturally subject to His dominion. All that He holds in his hand is under His dominion, from the crooked serpent in the heavens (Job 26:13) to the least hair on our head (Luke 12:7). Because we are creatures, God exerts over us ownership and supremacy. Such dominion imposes on us a natural duty towards God, an obligation resented by the one without love for God. The unbeliever does not appreciate that the immediate influence of God gives motion to his heart.

All the twentieth-century hubbub about human rights, or the recent tumult over rights of nature and animals, has all but drowned out the reality of Creator's rights. God's rights as Creator are fundamental. He further accentuates His dominion rights by the attention He shows to His creation, and through the gifts He gives to men and other creatures. We are like little birds with opened mouths and should not forget all His benefits. He provides food in season, rain on the earth, and innumerable tangible blessings. We owe Him everything.

In the beginning, before Adam's sin, the entire cosmos was a harmonious whole. Each part worked flawlessly with every other. Even after the fall of Adam, and the consequent marring of the Earth from the curse, the material creation retained qualities of beauty—God continues the shower of blessings. In Ecclesiastes 3:11 God reveals that "He has made everything beautiful in its time," so teaching us that it is the hand of God that is in all those changes surrounding us. He governs the gentle breezes blown on butterfly kisses and the consuming blasts of the furious hurricane.

The harmony of His works testifies that God rules well. But our eyes no longer see clearly, and there are temptations to complain against the providences of God. Sometimes His works are terrible to behold. As a ruler He governs everything that we see, whether we find it comforting or terrible.

## My Clay Pots

God paints His dominion rights over the creation in a stark mural with God as a potter and man as clay. Does the potter have the right to use the vessel in any way He wishes, even to smash it? In Romans 9:21–23, the battle is joined and the sword of the Spirit, the Word of God, strikes: "Does not the potter have power over the clay, from the same lump to make one vessel for honor and another for dishonor? What if God, wanting to show His wrath and to make His power known, endured with much longsuffering the vessels of wrath prepared for destruction, and that He might make known the riches of His glory on the vessels of mercy, which He had prepared beforehand for glory … ?" The answer to these rhetorical questions is a resounding Yes.

A few more verses underline God's utter sovereignty over every aspect of our lives. Nothing dare compete with His glory and authority, for "our God is in heaven; He does whatever He pleases" (Psalm 115:3). In another place says He, "Now see that I, even I, am He, and there is no God besides Me; I kill and I make alive; I wound and I heal; nor is there any who can deliver from My hand" (Deuteronomy 32:39). It is as futile to try to resist God as to call on idols for help. God will have mercy on all that call on Him in faith, and will crush with a rod of iron those that continue in willful obstinacy and sin (Psalm 2).

Because His throne is in heaven, and not on Earth or some other planet, there is nothing within creation to control God. His kingdom is literally out of this world (John 18:36). That is why Jesus, who is God, said of Satan (John 14:30) that he "has nothing in me." God is a power and law unto Himself. From Scripture we learn that God made all the Earth for His glory. He made us for His rule.

Although we have seen that God has a special place for mankind, yet we are still very much putty in His hand; He is king, we are subjects, and His dominion does not depend on our consent. Of great Pharaoh we are told: "For the Scripture says to the Pharaoh, 'For this very purpose I have raised you up, that I may show My power in you, and that My name may be declared in all the earth'" (Romans 9:17). God's dominion is not by the permission of His creatures. He owns us, we are His private property. And if we complain about this, He says, "But indeed, O man, who are you to reply against God? Will the thing formed say to him who formed it, 'Why have you made me like this?'" (Romans 9:20).

For all His power God is not capricious. He is most certainly a fearsome God, but He is not capricious, He does not change. His dominion over mankind is not greater or lesser in power now than before. Dominion does not adjust to conform to ever-changing human ideas of rights. As Malachi (3:6) wrote, "For I am the Lord, I do not change; therefore you are not consumed, O sons of Jacob." Though His dominion is irresistible, God is no human tyrant. He is all mercy, righteousness, and justice. This is His nature, and He does not change. He may be terrible in His awesome holiness, but He is also very good.

In His rule over body, but also over soul, His dominion is like no other. How then does the dominion that God exercises differ from that which He bequeaths to His offspring? There are dominions on Earth, orders and kingdoms even in the spiritual realms. God exalts His dominion above them all, says the Scripture, and He reigns forever, "far above all principality and power and might and dominion, and every name that is named, not only in this age but also in that which is to come" (Ephesians 1:21).

God owns us much as He owns the stars and dumb creatures. We are at His complete disposal, to use as He pleases. As we will see, He pleases to raise mankind almost to divinity, to sit at the left and right hand of God and around His throne. He pleases to place man at the crown of His creative work. No other creatures of the cosmos share this honor.

Indeed, God pleases to use the animals and all else to display His mercy to mankind. "What is man," says the Psalmist (Psalm 8:4–6), "that You are mindful of him, and the son of man that You visit him? For You have made him a little lower than the angels, and You have crowned him with glory and honor. You have made him to have dominion over the works of Your hands; You have put all things under his feet." The special love of God to man amazes the Psalmist, for he recognizes that our outward parts are in many ways inferior to the animals. We are not so hot as the sun, hard as the rocky hills, fast as the bird, strong as the ox.

Yet for all our weaknesses, and their strengths, beasts are almost like stage props in the history and work of the redemption of man. The work of redemption is the end of the work of creation, to the glory of Jesus Christ. This is the true central organizing principle at work in this world, not idolatrous environmentalism. Theologian Jonathan Edwards accurately summarized the state of things: "In one word, the greatest and last sight, which will be the issue of all events, at the winding up of things and dissolution of heaven and Earth, will be the complete accomplishment of the effect and end sought, by what Christ doth in the work of redemption."[348]

## Dominion, Not Domination

Everything is secondary to the glory of God in Christ. Therefore we do not have absolute property rights over our own persons. It is sin to harm ourselves unjustly because human beings are God's creatures. We are His possessions, as it says (Psalm 24:1): "The earth is the LORD's, and all its fullness, the world and those who dwell therein." Environmentalists imagine Christians think that man receives authority to crush the Earth. Not so, for the Earth and all creation is also the private property of God. If man is then granted possession of anything it is through an act of God's grace and not something to be sinfully abused.

There are people who intentionally abuse their power over the rest of creation, their fellow humans included. They assert in practice, if not in principle, the godhead of humanity. Scripture gives no ground for an elitist order where some men grind others into the ground and the rest of creation along with it. Such ungodly domination one should attack as fiercely as many environmentalists do Christianity. God did not give man authority to lord it over other men, but to rule over the Earth (Genesis 1:26–28).

Because we belong to God we must realize that He intends that we spend our lives for His glory. Our life does not have some sacred quality of its own that makes it worth protecting, preserving, and defending at all costs. The glory and truth of Jesus Christ must take precedence over our life. If we preserve our lives at the expense of God's truth, we walk the broad road that leads to the loss of our life and eternal soul. Indeed, Jesus warned, "whoever desires to save his life will lose it, but whoever loses his life for My sake will find it. For what profit is it to a man if he gains the whole world, and loses his own soul? Or what will a man give in exchange for his soul?" (Matthew 16:25–26).

"Or do you not know," exclaimed Paul the apostle (1 Corinthians 6:19–20), "that your body is the temple of the Holy Spirit who is in you, whom you have from God, and you are not your own? For you were bought at a price; therefore glorify God in your body and in your spirit, which are God's." How remarkable. If possession of our own bodies, with which we have unprecedented intimacy, is not an unqualified right, how then can anyone in their proper mind imagine that we have a limitless right to do as we please with the other creatures of God?

But the absence of an absolute right never suggests the absence of a duty to exercise dominion. God commands all mankind to exercise dominion over the creation. Dominion is not an option, it is a command. The ungodly at least recognize the fondness of men to lord it over creation, but they err in seeking to drive out distinctions between man and beast. Peter Singer, the perverse Princeton University philosopher, tries to neutralize this fondness by the environmentalist argument that "it is arbitrary to hold that only human beings are of intrinsic value."[349]

The problem for Singer is that he does not believe in true sin. Thus he really is the one who has no criterion other than himself to decide what is good and desirable in itself. This relativist morass leads him to question the motives of people who label as evil regimes they dislike: "But who is to decide what counts as an evil regime?"[350] Adolf Hitler floundered in the same muddy thinking, though Singer hedges his own position by saying "almost everyone would agree" that Hitler was evil.

Interestingly, in *Mein Kampf* Hitler attacked anthropocentrism and human efforts to master nature, writing, "Man's effort to build up something that contradicts the iron logic of Nature brings him into conflict with those principles to which he himself exclusively owes his own existence. By acting against the laws of Nature he prepares the way that leads to his ruin."[351] So Hitler found it good and desirable to impose environmental edicts on the German population, for example via the *Reich Agency for Nature Protection*.[352] He was himself an organic vegetarian, into holistic health care, would not touch meat, and a generation ahead in his other environmentalist sensibilities.[353]

Biehl and Staudenmaier's interesting little book, *Ecofascism: Lessons from the German Experience*, paints what should be a sobering picture for Christian environmentalists. Professions of love for nature, rather than being warm and fuzzy, can often be but the makeup on a monster; ecological concern is no guarantee of humane ethics or politics. History serves a warning that much of modern environmentalist ideology, despite ecological camouflage and professed principle of nature love, is fundamentally regressive and authoritarian, its ruling passions peevish and pitiless.

The urge to dominate others is not, however, unique to environmentalism; it is an unhappy human trait. It is the broad way of the world for men to lust for power over others. We all do it. Lust and envy are impossible to deny without divine help. Forget about ruling over pigs and dogs, for misanthropists the real action and excitement lies in the rule and dominance over people. Environmentalism just provides a contemporary pretext to satisfy the baser instinct.

One sees this urge to dominate even in little children, an urge that can be harmful or guided into fruitful avenues. It is not uncommon for a child to be domineering, to command total allegiance and instant obedience of a friend or sibling. "Give me the toy now or I will bash your head in" is less subtle perhaps than coercive environmentalist regulations, but the principle is the same. Greens have succeeded in getting laws passed that say, for instance, you will only flush with so much, and you may not use these light bulbs; to defy these infantile laws is to risk the wrath of Caesar. In Matthew 20:25–27, the account explains what Jesus thought of the common rule of ungodly men: "You know that the rulers of the Gentiles lord it over them, and those who are great exercise authority over them. Yet it shall not be so among you; but whoever desires to become great among you, let him be your servant. And whoever desires to be first among you, let him be your slave."

Regressive and unjust authoritarian behavior the Bible calls a sin. Such sin is different from Biblical dominion. "It is significant," according to theologian Rousas Rushdoony, "that godly man in Genesis 1:26–28 is called to exercise dominion (not domination) over the Earth, not other men, to develop all things in terms of God's law-word and to make this Earth into God's realm and domain. Fallen man does not seek dominion, which begins with his salvation and his ability to rule himself, but, rather, the goal of fallen man is domination, to control other people."[354]

The Christian way is to cultivate brotherly love and a servant spirit towards his fellowman. We respect the rights of his person and property. Private ownership of property—in other words, the right to dispose of it as the owner sees fit—is something towards which environmentalists (and Marxists) display much hostility. "Dominion," preached Ezra Stiles (1727–1795), former President of Yale College, "is founded in property."[355] Subordination of property rights to environmental dictates is a direct attack on the Biblical doctrine of dominion and a raw attempt to dominate other men.

Consider how Joshua, second in command to the prophet Moses, was called a minister (Exodus 24:13), thus one who serves. His role was to defend, not abuse, the rights of his people. The office of the priest, even for Aaron, was also classified as that of a servant (Exodus 28:1). In the church the rulers are called servants (Matthew 20:26), as are state officials (Romans 13:4). Such thinking has extended into the governments influenced strongly by Christian civilization. To this day, government officials in many countries are called public servants.

## Fruitfulness

The rule of God is always wise, but humans rarely manage their dominion by the rule of wisdom. Scripture speaks in Romans 11:33 of "the depth of the riches both of the wisdom and knowledge of God! How unsearchable are His judgments and His ways past finding out!" Kings of the earth favor pride, tyranny, and oppression. Also, "[t]he LORD is righteous in all His ways, gracious in all His works" (Psalm 145:17), but man is naturally perverse and wicked. The dominion of man is rarely, perhaps never, free of wicked possibilities to exploit and destroy.

When God made the creation, He declared it good and commanded man to fill the Earth and exercise dominion. God did not rescind His command when Adam sinned, but repeats and emphasizes it throughout the Scriptures. For instance, in Psalm 8:6: "You have made him to have dominion over the works of Your hands; You have put all things under his feet." Our sinful nature is thus no reason to excuse ourselves from exercising godly authority over the creation. It is the way of the coward to shirk dominion for fear of personal sinful tendencies.

Before a long journey a master, said Jesus, gave a money to his servant, with instructions to take that possession and work it. When he returned the master rebuked the wicked servant who had buried the money rather than use it to bring increase (Matthew 25:14–30). Fruitfulness and increase are one clear goal and blessing of the exercise of godly dominion. In Eden the man was commanded to take dominion and, as if to elaborate for dullards, the mode of dominion is further detailed

(Genesis 2:15): "Then the LORD God took the man and put him in the garden of Eden to tend and keep it." The creation, the Earth in particular, but even the Garden, was given into the hands of man that he should cultivate its fruitfulness and possess it.

One finds an analogy, imperfect but useful, in the relationship between husband and wife. It is not politically correct to point out that the Bible compares a husband's Biblical dominion over his wife to Christ's ministry and life sacrifice for the Church. Consider Ephesians 5:24: "Therefore, just as the church is subject to Christ, so let the wives be to their own husbands in everything." Notice the command to leadership by the husband and subjection by the wife is not in some things, which is already intolerable to false theologians, but in *every* thing. The force of the Scripture cannot be blunted; the husband exercises an awesome authority over his own wife. Her body also is his to possess, though the analogy is greater than imagined since his body belongs also to her (1 Corinthians 7:4). What is more, if they are both Christians the husband must remember she is a joint heir in Christ. What is unambiguous is that mankind does not have an option to exercise authority—dominion is the command of God, and one of the primary goals, as in the marriage relationship, is physical and spiritual fruitfulness.

Some husbands insist on replacing dominion with domination. They treat their wives with domineering disrespect, not as fellow heirs in the glories of Christ. Yet a man's sin does not abolish God's command. He is not the head of the wife because he wants to be the boss and wants to dominate. It is not what he does, but what he is, that settles the direction of dominion. Godly dominion is love, since "he who loves his wife loves himself. For no one ever hated his own flesh, but nourishes and cherishes it, just as the Lord does the church" (Ephesians 5:28–29).

A man who marries a woman is her federal head, as Adam was the head and representative of our entire race. Scripture declares it is not in Eve that all die, but in Adam because he was the head (1 Corinthians 15:22). Adam was to exercise godly dominion. His failure to do so did *not* absolve him of responsibility. A man's exercise of headship over his wife is to be for her fruitfulness and the glory of God. He is to deplore spiritual barrenness, in particular, and seek her good as a "fruitful vine" (Psalm 128:3). Children are one of many products of this fertile marriage, the fruit of the womb (Deuteronomy 28:53).

Earthy tones in this fundamental human bond are not a passing quirk. The principle of dominion seems plain. In 1 Corinthians 11:9 the Bible says, "Nor was man created for the woman, but woman for the man." So too the creation was made for mankind, not mankind for the creation. It is therefore no surprise for Jesus to teach that even time is subject to godly dominion: "The Sabbath was made for man, and not man for the Sabbath" (Mark 2:27). Of this dominion mandate Jonathan Edwards makes the following comment: "All of those are below him, and all … are made for him to be subservient to his use: their special end respects him, but his special end don't [sic] respect them."[356] Man was never to be on a par with the creation, but to be over it, yet under the Lord. The right of Gaia to her own children, the plants, the animals, the rocks and streams, and even the rights of these creatures as creatures, are so totally negated to be absent as even a factor in the drama.

## Fig Trees Multiply

God granted man dominion over the cosmos before the fall. Why did He not revoke it when Adam sinned? The answer lies in God's sovereign plan to glorify Himself in the salvation of fallen mankind. By His sovereignty He might have prevented any rebellion of angels, and so ensured the eternal obedience of Adam and his posterity. Yet He elected to allow Adam to sin. Does not the infinite grace of God in Jesus Christ shine the more brightly, as a many-faceted diamond, for its setting on the darkened backdrop of desperate wickedness?

Our fall has not revoked our authority over the rest of creation. Scripture says (Psalm 8:6–8), "You have made him to have dominion over the works of Your hands; You have put all things under his feet, all sheep and oxen—even the beasts of the field, the birds of the air, and the fish of the sea that pass through the paths of the seas." This scripture repeats the original Genesis 1 command to dominion. The fall has, however, changed the nature of man's dominion. It has made it a sweaty struggle. Before the fall man and the rest of nature functioned harmoniously. With Adam's fall came the curse of the entire cosmos. Now the land produces thistles without effort, but for figs it requires the sweat and thoughtful stewardship of man (Genesis

3:18; Matthew 7:16). Jesus taught by His example that fig trees gain value from their service to mankind. In one instance He cursed a fig tree without fruit, causing it to be forever barren (Matthew 21:18–22). He might as well have torn it up by the roots or poisoned it for the effect of this particular curse. Dominion demands fruitfulness; parts of creation that do not serve mankind are without value and fit only for burning. At the fullness of time, when God reveals His redeemed, even the Earth and the whole cosmos *will* melt and burn away (2 Peter 3:10–12) when their service is complete.

"A certain man had a fig tree planted in his vineyard," said Jesus, "and he came seeking fruit on it and found none. Then he said to the keeper of his vineyard, 'Look, for three years I have come seeking fruit on this fig tree and find none. Cut it down; why does it use up the ground?' But he answered and said to him, 'Sir, let it alone this year also, until I dig around it and fertilize it. And if it bears fruit, well. But if not, after that you can cut it down'" (Luke 13:6–9).

The Lord is speaking of much more than fig trees, but the greater lesson does not limit a less important lesson about fig trees. Men may govern and own fig trees, and men may destroy those that are not fruitful, and therefore do not serve man. Fig trees receive value in their utility to mankind. Since Adam the cursed creation no longer works towards natural support of human dominion. The ground must be worked, sometimes broken, sometimes destroyed, and in the various acts of dominion fruitfulness is produced. In taking dominion over the land care must be had to improve it. Land left to itself is not good. It was good for a brief time in Eden, but now it is not paradise. God cursed it.

So-called "natural" or wilderness areas are not hospitable to man, and God does not consider this a good or natural state. The presence of man is necessary to bless the land, "lest the land become desolate" (Exodus 23:29). The necessity of our dominion does not, however, mean that we use land indiscriminately. Instead, the owner of a fig tree must use it with a view to long-term fruitfulness and benefit for himself. To take another instance, in case of war, soldiers should not destroy fig trees, but need have no qualms of conscience about felling other trees with no obvious utility value for man. "[Y]ou shall not destroy its trees by wielding an ax against them; if you can eat of them, do not cut them down to use in the siege, for the tree of the field is man's food. … Only the trees which you know are not trees for food you may destroy and cut down" (Deuteronomy 20:19–20). God has made clear that fruitfulness proceeds from His hand, and He ordains that a second cause is the work of man in taking godly dominion of the whole cosmos. The new man in Christ, in particular, has a duty to exercise dominion in every area.

## Moaning and Groaning

God's kingdom undoubtedly reaches to all men and all the created order, but it brings salvation to none but the elect. God narrows the field still more, and sets His *special* care on His Church. They alone are the ones He calls the apple of His eye (Zechariah 2:8), His beloved (Psalm 60:5), His wife (Revelation 21:9), and many other terms of endearment. It was not for everyone that God became man, but for His Church alone. Jesus prayed, "I pray for them. I do not pray for the world but for those whom You have given Me, for they are Yours. And all Mine are Yours, and Yours are Mine, and I am glorified in them" (John 17:9–10).

For reasons of His own, reasons that boggle the mind, almighty God chooses to set His love exclusively on our kind and not on the other creatures. All we know is that, ultimately, it is for His glory. God chooses to glorify Himself through His work of redemption of the Church made up only of human beings. They share in His exaltation and will rise to reign with Him. The brute creatures are all instruments to effect this cosmic drama and are without intrinsic value in themselves. When they are valuable, their value is expressed by their utility in effecting the grand plan of man's redemption. Thus Jonathan Edwards wrote: "God sometimes made use of brute creatures as instruments to do his work. So he made use of two she-bears to punish the children that mocked Elisha. Sometimes he makes use of objects without life. So he made use of the stones he sent down from heaven to destroy the Amorites."[357] Fig trees, bears, hornets, rocks, and stars all work towards God's exquisite ends.

The creatures are thus tools of dominion. As godly men obey God and exercise dominion over all creation, through fruitfulness and multiplication, they glorify Him. That is why the future is bright. Much as the world and the Dragon may rage, they can never win. Jesus rules now, and evermore, and so the Church rules directly through His power and authority (1 Corinthians 15:24–25; 2 Timothy 2:12). John Calvin wrote, "Unless Christ were seated at his father's right hand, and had obtained supreme dominion, causing every knee to bend before him, the Church could never exercise its power."[358] Since we are products of His power, our dominion is just a branch of His. Human beings, and Christians in particular, in their authority over the cosmos "are governors by Divine providence; God is governor by his sole nature."[359]

Creation is designed for mankind, not we for it. If fits into God's redemptive plan for mankind, not we for it; we are the image of God, not the rest of creation. Nonetheless, in His great mercy God has not forgotten the rest of creation. God has a plan. He has ushered in what Jesus called the Kingdom of God, and we must occupy till He comes again (Luke 19:13).

In occupying, in exercising dominion, God steadily restores His blessing to mankind, and through mankind to the rest of creation. "The kingdom of God among men," says Calvin, "is nothing else than a restoration to a happy life …. It is true and everlasting happiness … 'newness of life' (Rom. 6.4)."[360] Far from being a destructive force to nature, as the Green Dragon says it is, Christianity is the only power that can restore nature. But redemption is not simply a return to the *status quo ante*. Christianity is not a utopian movement back to Eden, as the environmentalists would have it. Christ came to make a *new* creation.

Through the ages men have subjected the Earth to much abuse. It was forced, as it were, to open its mouth and swallow Abel's blood. It was terrible when Cain used splintered rays of sun to murder his brother. Even today the sun shines and acts as a servant to all manner of lusts and wickedness. The rain falls on the righteous and unrighteous alike; each day brute beasts and all parts of creation serve man's corruption. Little wonder creation groans in distress. "For we know," says Paul in Romans 8:22, "that the whole creation groans and labors with birth pangs together until now."

Rather than ending it all immediately, God elects to extend the bondage of the creation—through the curse of God and the sinful acts of men—

but only for a short time. Jonathan Edwards sets the context: "Were it not that so is the sovereign pleasure of God, the earth would not bear you one moment; for you are a burden to it; the creation groans with you; the creature is made subject to the bondage of your corruption, not willingly; the sun don't [sic] willingly shine upon you to give you light to serve sin and Satan; the earth don't willingly yield her increase to satisfy your lusts; nor is it willingly a stage for your wickedness to be acted upon; the air don't willingly serve you for breath to maintain the flame of life in your vitals, while you spend your life in the service of God's enemies. God's creatures are good, and were made for men to serve God with, and don't willingly subserve to any other purpose, and groan when they are abused to purposes so directly contrary to their nature and end."[361]

When John the Baptist preached the good news, and when Jesus came, the world enjoyed partial deliverance from this bondage. Jesus said (Mark 16:15), "Go into all the world and preach the gospel to every creature." What did Jesus mean when He spoke of every creature? He brings salvation to man alone, and teaches that we may not limit this offer to any class of men; we are, all of us, of one bloodline. And so, God offers salvation in Jesus Christ to "all men" (1 Timothy 2:4).

But what should one make of His reference to the creature? Puritan theologian Cotton Mather helps: "Through man and his religion, all creatures are joined to God, in order that their creation by God not be in vain."[362] And again he writes, "It is not for nothing that *Mankind* is in the *Gospel* called *every Creature*; he that beholds Man, may therein behold what is most wonderful in every Creature."[363] In other words man, as the head over creation, is responsible for it. "Indeed," writes Mather, "he that speaks to MAN, speaks to *every Creature*; and *Man* is therefore the more *concerned*, as well as *capable* to hear *every Creature* speaking to him."[364]

It is through godly dominion, through propagation of the Christian gospel, that creation begins restoration from the curse and bondage. But the final glorious day, when God makes all new, is not yet come. Even the dumb creation waits with expectation for that day of the Lord: "For the earnest expectation of the creation eagerly waits for the revealing of the sons of God" (Romans 8:19).

## Conclusion

Through godly dominion mankind aims to subdue the world, the flesh, and the devil, to take every thought captive to Christ. We are to do this, our faults notwithstanding, by being fruitful and multiplying, and by possessing the Earth, causing it also to be fruitful. Though the exercise of dominion by wicked men is cruel, this does not leave humanity with a valid excuse to shirk dominion responsibility. Nonetheless, our fatal history leaves little confidence in a happy outcome for anything under the dominion of man. That is why it is imperative to preach the gospel of Jesus, not environmentalism. Until the gospel has conquered the Earth, men will continue to be cruel towards other creatures. To follow the lore of the Green Dragon is worse than useless; it is to make geldings and then demand fruitfulness.

However distasteful this word "dominion" may be to the Green, it is a word associated inexorably with the Christian Bible, the source of all truth. How dare we contemptuously shunt it aside like a stale doughnut. If we are mature we are able to face unpleasant truths, and unpleasant truths we will face as long as we are in this world. Radical environmentalists note man's wickedness. But most do not believe he is hopeless, naturally insidious and vicious. If they believed this they would not seek to tie down the zombie Sin with Lilliputian ropes of education and regulation. It was not always like this, for God made man upright, but the natural man has now what the Bible calls a desperately wicked heart (Jeremiah 17:9). God's Word to mankind, accurately recorded in Genesis 1:26–28, says the following:

"Then God said, 'Let Us make man in Our image, according to Our likeness; let them have dominion over the fish of the sea, over the birds of the air, and over the cattle, over all the earth and over every creeping thing that creeps on the earth.' So God created man in His own image; in the image of God He created him; male and female He created them. Then God blessed them, and God said to them, 'Be fruitful and multiply; fill the earth and subdue it; have dominion over the fish of the sea, over the birds of the air, and over every living thing that moves on the earth.'"

This thinking revealed in the Bible, God's own, is anathema to the Green Dragon. It reasserts unequivocally what we saw in the previous chapters: man is separate from and exalted above the other creatures. So far from being divine, atoms, stars, gold, stone and all the inanimate creation receive not even an honorable mention. Human dominion over the creation drives environmentalists to apoplexy because it suggests a judgment of value, and mankind is at the pinnacle. Even before his creation man had a rich inheritance; all of creation is his to take and govern. God prepared the Earth as a beautiful home outfitted for man; indeed "the earth He has given to the children of men" (Psalm 115:16).

After God created the world, as a cosmic theater, He placed man in it to behold the marvelous works of God. God placed everything at the disposal of man, even the mighty angels to be ministering spirits (Hebrews 1:14), though he is not to lord it over other humans (Mark 10:42–43). Even after the fall, and His cursing of all the visible creation, God did not lessen man's dominion, but reiterated it.

Eternal death was wrought in the world, but God immediately adds the history of His restorations via godly dominion, mostly through the redemptive work of Christ and His Church. We have discovered how the Earth left to itself is *not* paradise, but a cursed and desolate wilderness that man is to improve and render fruitful to the glory of God and service of mankind. Yet even the wilderness brings great glory to God and is beautiful (Psalm 8).

Fallen man too keeps some beauty, though disfigured by sin. Given our tragic past we can have little hope in utopian human schemes aimed at subduing man or living in harmony with the Earth. Proper dominion requires a heart change, and this is God's work. Therefore the true Christian Church is indispensable in the work of restoring the Earth. Unless, and until, the true Church of Jesus Christ begins to behave as a servant and witness of Christ, there can be no hope of humanity living in peace. When nations adopt Christian positions they are no doubt blessed. For instance the nations that have respect for private property are uniformly more clean and prosperous than those with full-blown socialism. But this is never enough. To adopt Christian principles without a spiritual heart-change, as a mere means to material success, is an insult to God. Indeed, nations that embrace Christ and adopt the worship of God will be built in the midst of God's people (Jeremiah 12:16).

Human dominion is intimately wrapped up in the dominion that God Himself exercises. As Calvin wrote, "God would never allow [the Church] to be altogether destroyed, since upon the event of its destruction he would cease to be a king."[365] All of history spirals towards this one point, that God has kept the human race to prove His special care for His Church. There lies hope for the entire cosmos. Even the creation does not groan in vain, but in hopeful expectation. It is not a death spiral, but a life spiral. The dominion of Jesus Christ will steadily change hearts and cultures until "the earth shall be full of the knowledge of the LORD as the waters cover the sea" (Isaiah 11:9).

As the gospel conquers the world the wilderness must become cultivated and fruitful: "He shall cause to take root in Jacob; Israel shall blossom and bud, and fill the face of the world with fruit" (Isaiah 27:6); "the desert shall rejoice and blossom as the rose" (Isaiah 35:1); and "everyone shall sit under his vine and under his fig tree, and no one shall make them afraid" (Micah 4:4).

# CHAPTER SIX

## Death Wish: How the "Sustainability" Chimera is Harmful to the Environment and Politically Dangerous

*The real issue is not whether one cares about nature, but whether one cares about people. Environmental sympathies are not in dispute; because one puts the interests of one's children before the interests of the people down the street does not imply that one hates the neighbors, or even is disinterested in them. The central matters in dispute here are truth and liberty, versus the desire to impose one's aesthetic and moral tastes on others.*[366]

Julian Simon

## Earth Ocean

Water covers over seventy percent of the surface of our planet. Earth is its name, but ocean works just as well. All cant to the contrary, the Earth is still really a blue planet, not Green. Evidence suggests the past saw the whole globe covered with a thick warm blanket of water and balmy climes. In our day huge portions are near permanently frozen. In such a chilling environment ice begins to form on water and endures in some places like a stiff scab on the curved breast of the Earth. On frigid winter days the silence can be deafening. Deafening also is the insistence by environmentalists, including many professing Christians, that nature has intrinsic or inherent value, sacredness in and of itself, and therefore should be touched lightly, or not at all.[367] This dominant paradigm of "no change," like a mighty flood, threatens to drown the social, political and scientific debate." Its followers press increasingly heavy obligations on the mainline churches, whose progressive leadership enthusiastically encourage orgies of self-flagellation in service of the Green Dragon, "to set the example and be a light in this world greening our churches."[368] In this chapter we consider the implications of believing in a sacred Earth, a world in which man should be only a minor player.

First, note that environmentalism is an intolerant ideology. Freeman Dyson, a Princeton University physics genius, believes this is so because it is really "a worldwide secular religion."[369] The debate is over, we are told, and it is time to change the world in service of Green priorities; dissent is verboten, frozen out of polite conversation and characterized variously as "treason,"[370] "crimes against humanity,"[371] racist,[372] and similar epithets. Former European Union president Václav Klaus wrote, "Environmentalism is a political movement that originally began with the intent to protect the environment—a humble and perhaps even legitimate goal—but which has gradually transformed itself into an ideology that has almost nothing to do with nature."[373] Unbroken streams of half-truths, outright lies, and emotional manipulation drive environmentalist arguments, like raging waves of the sea, to the end that we must revere nature for its own sake irrespective of any benefit to man or any command of God.

But the hypothesis of *intrinsic* value is only significant relative to that which is unchanging, namely God, for He is the source of all value.

Having rejected the foundations for the rational calculation of value non-Christians who argue for the intrinsic value of nature are vain babblers; each person defining for himself the value of an object. Each man, being his own god, does what is right in his own eyes. And since our personal human preferences are notoriously fickle, value becomes a moving target, and then what happens to the idea of intrinsic value, poor thing?

The intrinsic value argument has a native attraction. It is almost mesmerizing, like a flickering fire. But like the flame, whatever changes always is hard to pin down. Professing Christian environmentalists do claim to believe in the Divinity. How then will they decide the value of a society, person, place, or object? Will their basis be the bare authority of human wisdom and changing personal preference, as it is for unbelievers? God's Word is the guide that Christians need if they want to be sure they rightly discern the value of nature. Without recourse to hearing what God has spoken, the professing Christian Green is no better off than the atheist environmentalist.

## Butterfly Wings

Everything humans do changes the environment and, from environmentalists' perspective, invariably for the worse. They believe that nature has value and goodness in itself. Therefore any gross human disruption of the environment is intrinsically immoral. It might be possible to ameliorate some of the damage we have done, but damage is inevitable. Our best human intentions inadvertently bring destruction; our gentlest exhalations, softer than the butterfly effect,[374] inevitably rain gales of destruction on the Earth. Yes, butterflies flap their wings too, but environmentalists do not blame them for causing hurricanes; they blame people.[375]

Clearly, concern about a healthy environment has progressed way beyond that which everyone in his right mind desires, such as reasonably clean air and water. Environmentalists have become so giddy with their newfound political clout that in their impatient zeal they would even ban elements on the periodic table, such as chlorine.[376] Starting in 2009 the *Supreme Court of the United States of America* and the *Environmental Protection Agency* formally declared carbon dioxide gas to be a pollutant.[377] By living we exhale the same; almost all living things do, and have done so for the history of the earth. Now, by government fiat, judges and bureaucrats define our breath as pollution, a phantom menace. But even trees emit the same gas nocturnally. And so, for part of the day, trees add to the planetary pollution problem. Some forests may even be *net* contributors to increasing atmospheric carbon dioxide.[378] Such uncomfortable possibilities leave us in a position where, if we wish to satisfy environmentalist expectations, we might need to cut down some fine forests to save the planet. What is a polite professing Christian environmentalist to do?

He or she could become a follower of the Green Dragon. Al Gore is gleeful at the prospect, writing, "Happily, it has recently become clear that a great movement to protect the earth is stirring deeply within the faith, and many religious leaders are now sounding the alarm."[379] Professor Paul Ehrlich saw the stirring decades ago, noting how "winds of change are clearly blowing in religion."[380] "In our world today," declares the National Council of Churches (NCC) for EcoJustice, "we are facing perhaps the greatest challenge ever put before us as a global community." The NCC claim is that our energy use causes global warming, which in turn "is fundamentally altering the world around us and faster than we can transform. We are damaging God's Creation."[381]

Methodist environmentalist Bill McKibben spreads the guilt trip on still more thickly, arguing, "We have changed the atmosphere, and thus we are changing the weather. By changing the weather, we make every spot on earth man-made and artificial." To be sure then the problem is not so much that we have done an evil to the creation, but that we have touched it at all. McKibben expounds this perspective: "We have deprived nature of its independence, and that is fatal to its meaning. Nature's independence *is* its meaning; without it there is nothing but us." For those who still fail to understand the hint, he adds another shove to help the guilt trip along: "[Y]ou will not know whether the temperature is what it is 'supposed' to be. Or if the sun feels sweet on the back of your neck, that's fine, but it isn't nature. A child born now will never know a natural summer.... Summer is going extinct, replaced by something else that will be called 'summer.'"[382] And it is all your fault. As shown by Robert Fowler[383] and others, environmentalists spend much energy to cultivate Green guilt into a regular dish of Christian culture; guilt-ridden sheep are more easily manipulated into becoming the main course.

The proper response to such manipulation is to burst into raucous laughter. We need not wade into water where the bottom is invisible and circling sharks sing sweet nothings. To be fair, some of the more consistent and thoughtful environmentalists, such as Paul Ehrlich[384] and E. O. Wilson,[385] do admit the reality and do not stoop to such shallow sophism. Ehrlich's name, after all means, "honest" in German. We were never meant to be independent of nature. It literally forms us, atom by atom, cell by cell, in our mother's womb. God uses it to feed and sustain our lives. The Earth marks us and, at death, our flesh returns to the Earth. Then as our spirits leave, and only then, do we become independent of the material creation.

But just as the Earth marks us, so we mark the Earth. This is most obvious in places like Greece, where amnesia is no excuse for failing to recognize the presence of man. Greece is a country remarkably rich in history and natural beauty, its soil fertilized by the labor of thousands of years. Each stone and deep shadow seems to cry, "There be men here." The visitor to Greece cannot shake the impression that every pebble has been touched by man and admired, carefully turned in the hand, like a sweet thoughtfully rolled on the tongue, and then set in its proper place. The Thessalian plains, out of which rise the famous monasteries of Meteora, boggle the mind. In places they roll from horizon to horizon, in every direction. Though the hand of man has touched it all not one blade of grass or ear of grain seems out of place.

Of course the hand of man has changed the wilderness, has tamed it. And the change is, for the most part, beautiful. Deliberate, even massive, change to the wilderness is not necessarily bad; it can be, in fact, good.

## Sky Ocean

How interesting that *all* environmentalists fret over humans affecting the cosmos. In their creation myth life, man included, sprang from an ocean of prebiotic soup. But ages before slime-dripping man leaped at Earth's throat, fangs bared to tear and break her, there were bacteria.[386] The early atmosphere, so goes the myth, was virtually free of oxygen.

Oxygen was thus a pollution that killed the earliest organisms, as dangerous to them as the absence of oxygen is to their human posterity. Yet, according to the myth, microbes in the primordial soup slowly caused increasing oxygen concentrations to bubble up into the atmosphere, polluting the native environment.

Had the United States *Supreme Court* been around billions of years ago it might have ruled that oxygen is pollution. Presently paleo-microbes poisoned the sky with their oxygen "flatulence," much in the same way livestock are today considered the number one destroyer of the environment, at least according to the United Nations, for their forbidden emissions.[387] What is more, organically raised cows put out 16 percent more greenhouse gases than their non-organic counterparts,[388] meaning that organic is not obviously better. Of course, the protobacteria were also presumably organic, and we hear they were accidental genocidal maniacs. Mass extinctions, biospheric upheaval, followed the ancient atmospheric catastrophe and led to a wholly changed natural order. Strangely enough, environmentalists never seem to consider these biological disasters unnatural, or bad, or sinful, but "very good." In this sense, environmentalists do not have a genuine biological perspective on the Earth. They speak of humans as if we were alien or unnatural.

How can greens escape their own logic? Other creatures poisoned the atmosphere, but supposedly this is natural and part of the evolutionary path, certainly not immoral or wicked. When, so they claim, man changes the land or sky it is immoral and wicked, even though we and bacterial scum are relatives.[389] Perish the thought that man should act as an unreasoning animal or microbe! Yet when man uses reason to exercise power over nature, whether careless or thoughtful, he is charged with irremediable wickedness.

What gives? What gives is that the Green Dragon would be well content with the withering away of Christian religion, a silencing of the Biblical view that man is a special creature, spiritually sensitive to his closeness to God and cognizant of his lordship over the rest of creation. This Green irrationalism serves as a point of contact with the increasingly unreasonable mainline Christianity. And the Dragon is content to use the self-proclaimed weaknesses of the Church against its mortal enemy; if the Church wishes to make its own hanging rope, the Dragon will not stop her.

## Stewardship

Everything humans do has a ripple effect to the rest of creation, so there must be a careful accounting of the cost of our actions. Before a man begins to build he should carefully count the cost. This is a lesson with spiritual significance (Luke 14:28). It implies that economic development is not necessarily always good, but can be negative. In other words, environmentalist critiques or prescriptions are not necessarily wrongheaded in every instance. Criticism of rampant consumerism, for instance, is quite appropriate. Historian Gary North writes, "This [twentieth] century has been a radical historical aberration: large-scale mass production, financed by monetary inflation, accompanied by mass pollution, compounding yearly, decade after decade. This is not the culmination of Christian orthodoxy but of arrogant secular humanism which is steadily consuming its moral foundation, namely, the cultural veneer of Christian orthodoxy."[390] Jesus warns against greed and covetousness, reminding that "one's life does not consist in the abundance of the things he possesses" (Luke 12:15).

Modern high-pressure advertising seeks to define human beings in terms of possessions, what they consume, what they do and enjoy, and much of the culture buys it. These are the things the pagans "seek" (Matthew 6:32). Christians should seek to work diligently, to increase and prosper, but when they indulge their lusts to covetousness or gluttony they will not receive God's blessing (Hosea 4:10). God, in His providence, gives His people property, and He expects them to be fruitful and multiply, to be diligent and bring increase in whatever area they labor, and to enjoy the fruits of their labor. And if we are materially wealthy God warns us "not to be haughty, nor to trust in uncertain riches but in the living God, who gives us richly all things to enjoy" (1 Timothy 6:17).

Voluntary pursuit of poverty is no solution to gluttony or avarice. Indeed, poverty is not good, it is the ruin of the poor (Proverbs 10:15). In every instance righteousness is better, as it is written (Proverbs 28:6): "Better is the poor who walks in his integrity than one perverse in his ways, though he be rich." Coerced central economies also offer no help. They are merely the ugly side of the humanist coin. Even the rosiest apologists for the "scientific" socialist experiments in China, North Korea, Cuba, or the former Union of Soviet Socialist Republics

(USSR) must allow of a grim and continuing tale of devastation to both people and land, as environmentalists admit.[391] Socialist governments sought economic development, but their policies resulted in most of the famine deaths of the twentieth century, all in peacetime.

When we look at the present state of the world it is incontestable that pollution is also horrendous in many of the poorest countries with the lowest levels of political and economic freedom. By contrast, countries with the greatest levels of political and economic liberty tend to be the cleanest and the wealthiest. It was not always that way. During the early years of the Industrial Revolution, London was toxic and the great cities of Europe and the United States of America were not much better. Seoul, South Korea, of 15 years ago had air stale and brown as lukewarm coffee. By 2010, national energy use has almost tripled, the economy has swelled almost eightfold, yet the air quality is enormously improved and continues to improve. Like Japan this is now a relatively wealthy country with almost no natural resources, except the human resource.

Conversely, North Korea, with more abundant natural resources, began its socialist odyssey in a far more favorable position than the war-devastated south. Yet the North suffers disproportionate environmental pollution. An enslaved and increasingly impoverished people cause massive deforestation in efforts to eke out an existence in socialist utopia.[392] One gulag survivor, an escapee of Yodok, wrote of children in the concentration camp, "We ate anything that moved.... By the time a group of prisoners finished working a field, no animal was left alive. Even earthworms were fair game."[393]

Though it enjoys the support of the U.N. Secretary-General (appointed in 2007), Ban-Ki Moon, South Korea's radical Green movement is still nascent, largely a foreign import that is far from grassroots. This, plus the dramatic improvements they have achieved in pollution control, teaches another lesson: it is possible to have a reasonable policy of environment protection without making obeisance to the "environmentalism" of the Green Dragon.

The same trends are true in other countries where government greatly limits interventions in economic and industrial development. In fact, by now many studies have shown that as people become more affluent they begin to show great concern about a clean environment. As economic development reaches a level where humans have the liberty to care about clean air and other markers of a healthy environment, a liberty purchased by increasing economic development, pollution begins to decline.[394] In the words of Indur Goklany, "wealth and health reinforce each other in a virtuous cycle of progress."[395]

A lesson to glean from the Korean example is that poor countries are not poor because of pollution and environmental damage, but polluted and otherwise damaged because of poverty. Wealth and economic development usher in an improved environmental quality. Thus anyone concerned about an improved natural environment will encourage economic development. But economic development is precisely the change that Greens and their professing Christian friends despise; they characterize it as raw exploitation.

In *Respecting Creation's Integrity*, Calvin DeWitt argues that we humans are becoming an unbearable burden on the Earth and should cease to "exploit creation beyond meeting [y]our basic needs."[396] Elsewhere he suggests that when it comes to the Earth we should look, but not touch,[397] because the Earth is a rare work of art, like a Rembrandt. DeWitt seems to suggest the value of a Rembrandt is intrinsic and therefore nearly constant for all, when in fact the value of a work of art depends on quality, rarity, and many other historical or personal aesthetic considerations. The value of the living green creature first used to make the Rembrandt canvas was not intrinsic to its existence but depended on considerations such as where, when, and why it was planted. The tree could have been made into toilet paper. And probably DeWitt would reject association of art works with toilet paper.

Books on preaching get in on the act too, advising church teachers to weave every environmental shibboleth into sermons because, as one preaching manual opines, "concern for the environment is not an American issue, or a British issue, or a European issue, or a Japanese issue; it is a world issue. Children learn about the environment in school. They learn about the delicate balance of nature. … Our children are growing up with a knowledge that natural resources will not last forever. The world in which we live is the only world we have; we must take care of it. Children learn about recycling, and they teach their families how to shop for environmentally friendly products. Americans who once engaged in rampant consumerism are taking stock of their behavior because of environmental threats. We realize that what we do affects others."[398]

Overconsumption is thus proposed as a physical cause of the imagined ecopalypse. The solution offered is to live simpler and discover the spiritual benefits of poverty. This is like the crazy thinking of some Christians who pray for persecution because it marvelously focuses the mind on what matters most in life. Despite being the country environmentalists love to hate, overall pollution in the USA is dramatically lower today than just a few decades ago, and continues a downward trajectory.[399] What is more, as in South Korea, the U.S. environmentalist movement cannot claim much credit for these improvements, since the improving trends began *decades* before the magical 1960s, thus long before radical environmentalists came to have serious social and political clout.

## Kindness

Godly dominion is what we should aim for. Godly dominion should be vastly different from that practiced by the ungodly. According to theologian Adam Clarke, "The wicked, influenced by Satan, can show no other disposition than what is in their master. If they appear at anytime merciful, it is a cloak which they use to cover purposes of cruelty."[400] The compassion of the wicked is a tyranny and cruelty, not only towards humans but animals too.

Judeo-Christian culture places great emphasis on kindness to animals, and we remember that God honored the ox, as a representative of all beasts, to be close by at Christ's birth in its manger. By extension, we recognize that, like God, we must also delight in the creation. But we must balance this with another truth—the relative value of life is not equal. God made all the cosmos for man and gave all into his hand (Genesis 9:2). Time itself, which seemingly waits for no man, even it is a servant of sorts. Therefore when we do not use the creation in a fashion that contributes to human welfare we are behaving unjustly.

Still another factor must be considered. When Adam received authority over creation he became morally responsible for relating to the creation in a way that glorifies God. That men have rarely done so does not negate their responsibility to seek fruitfulness in dominion, and thus to *change* the world. But we must remember that our stewardship of the world, and its improving fruitfulness, is a secondary benefit that does not lessen human authority.

Scripture shows that we may justly draw not only physical, but also symbolic and spiritual benefit from the creation. We may, for instance, destroy creatures for food (2 Peter 2:12, Acts 10:13–15) and other utilitarian purposes, but also for science to increase knowledge (Matthew 24:32), for comfort (John 1:48), and most notoriously to cover shame (Genesis 3:7). In short, man is Lord of creation and may dispose of it at his pleasure, provided in doing so he violates none of God's laws.

Even untouched wilderness may have value to man. Jesus often climbed mountains to pray. Moses saw God's back on a mountain (Exodus 33:23), Peter saw Jesus' glory on a mountain (Mark 9:2), and the holy city sits on the spine of a mountain (Joel 2:1). But for all His attraction to mountains, wild and tamed, Jesus would flatten them to plains (Zechariah 4:7) or cast them into the sea (Mark 11:23) should their continued existence not fit with His plan of redemption for the Church. From God's example we learn how dominion encompasses the sobering power to destroy (Genesis 7:21–24; Matthew 21:19–20).

Professing Christian environmentalists say we should treat the Earth as God does. Though this is not possible, our nature and powers being so much lower than His, yet one can appreciate the sentiment. What they mean, however, is not actually so orthodox. They mean that we should step back and admire the divine handiwork, otherwise leaving it well alone. A problem with this is that it implies a clockwork God who is passive in upholding the world, who lets it run along with minimal interference. God is infinitely far from the environmentalist ideal as time and again He treats the creation as a mere tool to teach lessons to his people. In war He commanded His people to be brave against enemies (Joshua 11:6) and in battle to "hamstring their horses and burn their chariots with fire." Cutting the sinews of the horses would make them painfully lame, and one can hope that at least a few were soon put out of their misery and killed for the meat wagon. God's command made it good and necessary to do what in other circumstances would be sinful cruelty. Egypt's biodiversity He devastated in multiple instances, and the biosphere, or ecosphere,[401] of many other nations too. It is clear from Scripture that God does not consider the value of the creature to be a constant. In other words, their value is not *intrinsic*.

God brings the scourge on man and beast and makes lands to "tremble and sorrow" (Jeremiah 51:29). And why? All so humans might see and understand the display of His power (Exodus 9:16). All so redemption of man might be accomplished and applied. What enormous destroying forces He brought into play to flood the whole Earth and quench life (2 Peter 3:6)! Or what cosmic blasts accompanied the day He made the universe tremble and the sun stand still (Joshua 10:13)! Whatever the cataclysmic effects elsewhere in the universe—and consideration of star-stopping physics suggests that possibility—they are not mentioned. These omissions are not an indication that we are incapable of appreciating the Joules and Watts of dynamic destruction. God omits to give us the details because these effects on the redemptive theater are relatively unimportant in the grand scheme of things.

## Mercy

After heavy rains half-drowned worms lie panting in the overworld to arouse an ounce of pity in the heart of a man who seeks for the worm a dryer patch to burrow and live. How much more would the passionate screams of houghed horses arouse him, their eyes rolling madly in exquisite agony and helpless frenzy! But forget, for a moment, trees, worms, and horses. Think for a moment of all the mass of humanity that God rules with a rod of iron (Psalm 47:2; 66:5; Revelation 2:27). "I kill and I make alive," says He (Deuteronomy 32:39); "I wound and I heal; nor is there any who can deliver from My hand." How terrible God is! All flesh should tremble. He is no tame lion of Judah He.

"What shall we say then?" asked the Apostle Paul (Romans 9:14) about God's destruction of men made in His own image. "Is there unrighteousness with God? Certainly not!" If God has mercy on whom He will have mercy, and compassion on whom compassion, then how can we wonder that He destroys the stage and props of redemption when He judges necessary? Egypt's Pharaoh was no mere prop, but a player in his own tragedy. God raised Pharaoh to prominence on the stage that we might learn by his devastating fall, "that I may show My power in you" (Romans 9:17). We should know, if we are Christians, that God is no bland middle-of-the-road milksop fellow who just wants to get along with everyone, and leave their lives unchanged. God will do whatever it takes to redeem elect humanity, even so far as allowing the murder of His only begotten Son. God is the best friend to repentant sinners; He is love, and engraves them on the palm of His hand (1 John 4:16; Isaiah 49:16); they are never out of His sight, and He holds them close to His heart (Isaiah 40:11). But to the wicked all that remains is "a certain fearful expectation of judgment, and fiery indignation which will devour the adversaries" (Hebrews 10:27). In all these things we see, as He intends us to see, both His goodness and severity (Romans 11:22).

Some might consider it cruel to kill an animal by slitting its throat, or to come near it with a knife. It is certainly not the modern way. Yet such was the way God *commanded* the ancient priests carry out certain sacrifices. It was not an option, but was the prescribed mode so long as God's people practiced Temple worship. Modern sensibilities are clearly not true guides to discover what is, and is not, the fitting way to relate to the rest of creation; the more sure source of truth, namely the Bible, should be preferred. And the general tenor of the Bible verses so far referenced points out that we humans have considerable liberty to dispose of the creation at our pleasure.

## Balance

But, in the words of Spider-Man, with great power comes great responsibility.[402] Do we not fear we may disturb the balance of nature? If we do, who will save the planet? Enter the Green goblin. Rachel Carson, the author credited with introducing into the modern psyche the idea of a "balance of nature," pictures a man perched precariously on a cliff face, spider-like. He must pay careful attention to the gravity of his position, or kiss the dirt.[403] Carson's *Silent Spring* and the environmentalist agitation following its publication led, in the United States, to the 1972 banning of DDT, a miracle insect repellent that had virtually wiped out malaria in the West.

African countries have found themselves faced with malaria upsurges as they gradually eliminated use of DDT due to pressure from environmentalist and international aid organizations. The use of DDT in Mozambique, noted the *British Medical Journal*, "was stopped several decades ago, because 80% of the country's health budget came from donor funds, and donors refused to allow the use of DDT."[404]

Carson argued that it was unacceptable human arrogance to risk change to the ecological balance by disturbing the mosquito. "On this environmentalist premise," wrote Keith Lockitch in his analysis of Carson's work, "the proper attitude to nature is not to seek to improve it for human benefit, but to show 'humility' before its 'vast forces' and leave it alone. We should seek, Carson wrote, not to eliminate malarial mosquitoes with pesticides, but to find instead 'a reasonable accommodation between the insect hordes and ourselves.' If the untouched, 'natural' state is one in which millions contract deadly diseases, so be it."[405] Here again we see the deep idolatrous current of reverence for nature, the same as Heidegger's "letting-be,"[406] Wilson's urge to "leave alone,"[407] Kohák's need to look "a cow in the eye with a clear conscience,"[408] McFague's insistence that our reflexive impulse should be "to abstain,"[409] and Fox's insistence that it is sinful sadism "to get pleasure from the killing."[410]

Rather than waiting patiently for the gifts of Gaia, the Green accusation is that humans force the Earth, hence their accusations of rape and exploitation. Instead of waiting in humble dependence, humans irrigate. Instead of waiting for food to drop off trees into their mouths, they cultivate. Humans engage the planet with great goings-on. Rather than being passive, Western Civilization demonstrates the power of man as a creator, and thus as a changer of the environment. Radical environmentalists prefer humans approach Gaia on her own terms, receiving food and life when she chooses to give it … or not.

Some environmentalists allow in theory that the Third World has a right to develop economically, but as DDT has shown, they will not allow it, except on utopian terms that are difficult for even developed nations to achieve. It is like saying the people of Africa and India are welcome to eat, just that it should be caviar and champagne, not meat or wheat. "As far as food is concerned," writes Peter Singer, "the great extravagance is not caviar or truffles, but beef, pork, and poultry."[411] How convenient—caviar eaters can slay all the sturgeon babies they wish and fly on private jets guilt free. There is a carbon credit, or some such penance, for those with cash to burn. But those in the Third World who would like a mouthful of beef, pork, or poultry, they are the real gluttons and world destroyers! Let them eat cake; let them run their cars on biofuel made from wine, like Britain's Prince Charles;[412] let them eat caviar.

Ecological balance, or the pretended pursuit of it, is a pillar of environmentalist thought, one with deadly outcomes. In his scientific addendum to the novel *State of Fear*, Michael Crichton describes the DDT ban as "arguably the greatest tragedy of the 20th century." He sums up Rachel Carson's legacy as follows: "since the ban, two million people a year have died unnecessarily from malaria, mostly children. The ban has caused more than fifty million needless deaths. Banning DDT killed more people than Hitler."[413]

Ecophilosopher Erazim Kohák writes, "If anything could be said to be 'natural,' it would more likely be stability or the quest for it."[414] Similarly in his book, whose title begs the question, Al Gore writes, "If we try to adapt to the changes we are causing rather than prevent them in the first place, have we made an appropriate choice?"[415] Not surprisingly his answer is an emphatic negative. He calls instead for a change in human nature and behavior through reeducation or other means. This echoes other environmentalists, such as Christopher Manes, who wrote, "Learning to reinhabit the world, to restore a culture in balance with nature, will not be easy."[416]

It is not just media, politicians, and teachers pushing hard. By now corporations want to fly high on the Green breeze. *Delta Airlines* has turned its in-flight magazine into a Green rag, with almost every edition now carrying some tiresome story about ecopalypse. The April 2009 edition featured a discussion with Earth Day co-founder Denis Hayes, who longs for a national program with the motto, "Ask not what the Earth can do for you, but what you can do for the Earth."[417] Their April 2010 edition was entitled "Mother Earth."[418] The June 2009 issue of Air France's in-flight magazine repeats the usual Green canards: "[S]ince adopting a sedentary lifestyle 11,000 years ago, we have upset the balance maintained since life first appeared 3.8 billion years ago. The earth and man are linked in life and death; damaging it and exploiting it beyond what it can bear means imperiling the planet and humanity along with it."[419] Though not so damning in tone one finds a similar trope on international routes with *Korean Air*: "Global warming's impact on our planet has grown noticeably in recent years, and the

Earth's rising temperatures have meddled with the ecosystem."[420] Presumably this is why Going Green is so important, as the cover of the *US Airways* magazine[421] proclaims in April 2010.

These airlines have simply taken and uncritically repeated radical environmentalist talking points. There is dark magic in bold lies. Never mind that satellite-measured global mean surface temperatures—our most accurate and unambiguous scientific measurements—show that as of 2009 there has been no global warming for eight years in a row, and dramatic cooling the past two years.[422]

Though they add to the drumbeat it is of course not fair to accuse businesses of outright fabrication. They are just repeating what they have heard from politicians and some hysterical scientists. They are also hoping to ingratiate themselves with lawmakers who have the power to employ government controls in a way that forces consumers to use the products of their pets. There is a killing to make when government socialism forces large populations to use certain products, such as Green wind farms, rather than allowing free and uncoerced choices. Like a whirlpool, stampedes are reflexively and terrifyingly attractive. Various businesses and labor unions are increasingly looking to use the massive increase of environmental legislation and regulation to tie consumers and competitors in Green tape. Philip Mattera of the labor organization *Good Jobs First* says, "We're at a turning point that will have an impact on the future of the whole economy, and a lot of unions are gearing up."[423] It is only logical that large labor unions and businesses are trying to position themselves most favorably to make a profit and eliminate competition.

## Saving the World by Destroying It

The idea of collapse is predicated on the idea of balance, or equilibrium. Both are inflexible axioms in the environmentalist movement. The environmentalist believes passionately the hypothesis that there exists a natural balance that humans must preserve at all costs. Adjusting to or creating a new "balance" is neither proper, possible, or natural. Therefore, as far as is possible, we must try to keep everything the same from day-to-day. This is a blinkered and unrealistic view of the world, but environmentalists are not shy about imposing it on others. They are content to destroy what they have decided is "unnatural."

For instance, their philosophy moves them to destroy trees when they are "alien" species.[424]

Poor little Knut, the fluffy baby polar bear rejected by his mother, also faced rejection from the Green movement. In the wild he would have died of exposure or been eaten by one of his parents. A Berlin zookeeper had mercy on his plight and nursed Knut to health. People for the Ethical Treatment of Animals (PETA) objected, declaring that man's interference to save the bear was an offense against nature, something unnatural. What this meant, quite simply, was that Knut must die, for die he would if not fed by man. PETA spokesman Frank Albrecht said, "Hand-rearing a polar bear is not appropriate and is a serious violation of animal rights."[425] You see, to the consistent environmentalist preserving the sacred integrity of the natural world is an imperative.

Environmentalists lost that battle, and most have moved on. But had they won, Knut lovers could at least console themselves that wild polar bear numbers have risen 25 percent in the past decade. Greens would destroy nature if its precious—can we call it "personhood?"—is in any way directly influenced by man. The influence of man causes it to become inauthentic, and it appears that for environmental groups an inauthentic life is not worth living.

Václav Klaus sets this Green foolishness nicely into context: "The environmentalists seem to neglect the fact that a major part of the Earth is the result of conscious human activity and that the disputes the self-appointed protectors of nature so often initiate do not aim at protecting any original landscape but rather a historical product of human activities."[426] This reality means that a landowner need not experience existential angst because the decision to clear a stand of trees disrupts the mythical balance of nature. There are always costs in removing the trees, but one should weigh these against benefits rather than sifting both through the teeth of an invented and materially nonexistent Green ideal.

Of course, environmentalists are not dummies. They insist that slow evolutionary change has happened in the past, and must continue; they are all for hope and change. But they insist that since the Earth is ancient, some 4.6 billion years in the making, changes must take place at a glacial pace, most especially changes introduced by humans. According to Jeremy Rifkin, "The United States and other nations are consuming the earth's endowment faster than the biosphere can absorb and recycle the polluted waste and replenish the stock of natural resources our children's generation will need to sustain their economic future."[427]

There you have it. The positive rate of change of human industry and human population is, in the opinion[428] of environmentalists, problematic. Too rapid a change and nature becomes unbalanced. In this view environmentalists are deeply conservative, since the only hope and change in nature they are willing to tolerate must be almost imperceptible on the timescale of a human life. This is not an empirical scientific position, but one arrived at via the vagaries of ecological philosophy and boundless blind faith.

Because of their faith in this idea of balance, followers of the Green Dragon idealize primitive societies, who supposedly left no more impression on the land than a pack of animals wandering through the wilderness. All of these are patently absurd positions that some, not all, Greens continue to embrace, presumably for didactic purposes. On the one hand, according to Stanley Diamond, "Primitive society may be regarded as a system in equilibrium, spinning kaleidoscopically on its axis, but at a relatively fixed point." It is stable, and balanced. On the other hand, they say modern Western civilization is always intrinsically unstable. "Civilization," avers Diamond, "may be regarded as a system in internal disequilibrium; technology or ideology or social organization are always out of joint with each other."[429]

## Groves and Graves

I may like a certain collection of plants in my garden but should I impose my aesthetic preferences on everyone else? Perhaps I have the ear of politicians. Should I use this influence to have the police force my neighbor to craft a Japanese garden that I like? Radical environmentalists are not so demure and urge the immediate use of state power to impose their preferences on society.

Perhaps the average member of a Greenpeace, World Wildlife Fund, or some such group is no misanthropist, intent on making his neighbor suffer for the sake of the planet. But if this is not so, it is surely true of the leadership in the environmentalist movement, and is the formal position of most of these Green gangs. Ignorant members bear some responsibility for the way their donations serve the Green Dragon.

The Green Dragon encourages a backward looking perspective, to an idealized past when everything in nature was purportedly in balance. Precisely what this balance should look like is unclear, since all parts of nature, like climate, are always changing. Where does one find the mythical sweet spot? What types of plant and animal species should settle in a particular geography? In terms of a comfortable temperature in the thermosphere, the lower part of our atmosphere, what temperature is not too hot, not too cold, just right?

This environmental Goldilocks is clearly a fairy tale, far from the scientific reality. The Biblical ideal not only recognizes that change is unavoidable, but positively requires that man steadily transform his environment. Though ensconced in environmentalist history, the idea of balance in nature is also no longer a serious part of scientific debate. Václav Klaus makes an incisive observation, "Nothing like a predefined optimal state of the world exists that we should, for some reason, preserve and protect."[430]

Consider the tanned belly of Greece with its *wild* abandon, golden crops waving like a laughing Nymph tossing flaxen hair at the sun. Unlike Greece, the lush fields just outside Paris have a manicure marked by sophisticated precision. It is questionable, to say the least, that mythical ideas about a balance should have made the ancient Greeks concerned that grazing sheep and goats would change the land, or the ancient Franks resist the logging of virgin forests. Historian Edward Gibbon notes how "almost all the flowers, fruits and herbs which adorn European gardens were imported from foreign climates."[431] Is Europe therefore less capable, or maybe incapable, of natural beauty? This analysis seems quite unreasonable, but is exactly the dead end to which environmentalism leads.

From the environmentalist perspective human industry caused massive ecological destruction and was thus bad. Even worse, in the language of Methodist environmentalist Bill McKibben, Greek and French children will never know anything natural; they may think the passing sunset beautiful, but it is not, for man has changed the horizon and so deprived the sunset of its intrinsic dignity in passing away. Yes, the Greeks and French transformed the land. So what? One has permission to disagree with the silliness of environmentalist dogma and could argue that the new landscape is "more than an acceptable substitute—and not just from an aesthetic point of view."[432]

## Changing the Balance

Scriptures such as Isaiah 41:19–20 make plain how unbalanced is the propaganda of the Green Dragon: "I will plant in the wilderness the cedar and the acacia tree, the myrtle and the oil tree; I will set in the desert the cypress tree and the pine and the box tree together, that they may see and know, and consider and understand together, that the hand of the LORD has done this, and the Holy One of Israel has created it." The idea of ecological balance as a central organizing principle, keeping everything in relative stasis, is not a matter of urgency to the King of Heaven.

God speaks here and shows not only complete indifference to the "rights" of the creatures in the wilderness in question, but positively revels in His work of transformation. And let us be clear that His work brings devastation to the desert via invading shady tree species that environmentalists would consider alien. Here is an ecosystem that existed who knows how long, and God comes along to plant different trees that will change the water cycle, the soil, and air in a geological instant. He takes the desert, with its unique and fragile ecosystem, and undermines "balance" in acts that would bring species extinction.

In God's calculus it is sometimes good to change the bioculture of an area, often rapidly. Some environmentalists have called it genocide, but God declares such actions legitimate and commends to our consideration that He did it for the sake of human observation and education. It will be useful to quote extensively from theologian Jonathan Edwards, who speaks so clearly to this point: "God will cause waters to flow in the dry desert, and turn the barren wilderness into a pleasant garden, and bring forth these glorious things in such places that were most likely and most distant from any thing of this nature, and where was the least of any previous disposition, preparation, or foundation for any thing of this, that his hand might be the more visible, and his power and efficacy the more apparent and undeniable;

as of old, when God furnished a table in the wilderness, and brought forth water out of the flinty rock. Therefore here it is said, 'The Holy One of Israel hath created it.'"[433]

Scripture teaches that there is nothing intrinsically wrong with modifying an ecosystem, creating what environmentalists might characterize as an imbalance by addition. King David and his son Solomon, for instance, caused massive logging from the forests of Lebanon, taking trees without number (2 Chronicles 2:9). Abraham planted a grove of trees alien to a local ecosystem (Genesis 21:33). But often when the Bible speaks of trees it is to mark God's command to *destroy* ancient groves that pagans proclaimed sacred (Exodus 34:13; Judges 6:25; 2 Chronicles 15:16)—an imbalance by subtraction.

Sometimes God changes the world slowly. Other times the rate of change is rapid. In extreme cases a hurricane, earthquake, or even a global flood convulses all before it. Leaving nature well alone is clearly not part of Biblical pattern, not even for the most ancient society of all, that of Eden. Before his fall, Adam tended the Garden of Eden (Genesis 2:8–15). In paradise he was to dress and keep it. It did not have thorns or thistles, but left to itself it would still become wilderness, like the Earth outside the garden. Even the perfect garden was not complete in itself, but needed the wise care of man to keep it from growing wild. Wilderness was outside the garden, but at this time wilderness was still intrinsically good, in fact "very good" (Genesis 1:31). In the garden of Eden, in its pristine original, God showed Adam an example of how he was to cultivate the land as his posterity overspread the earth and steadily changed the wilderness into a garden.

With the fall came Adam's expulsion from the garden and the curse of the earth. The world was no longer a safe place of peace and plenty, but a perilous place where poverty, natural disasters, disease, and famine would haunt his posterity. Like a wild beast driven into an animal blind, God subjected the earth to maddening frustration; the wilderness today is not the way it should be; it is cursed. It is no garden of Eden, and can never again be such. "[T]he ruin of man," wrote theologian John Calvin, "drives headlong all those creatures which were formed for his sake, and had been made subject to him."[434] The curse notwithstanding, an examination of the breadth of history shows

His tender care for the human race, culminating in the resurrection of Jesus Christ. Calvin is helpful again, writing, "The end to which the whole scope of the history tends is to this point, that the human race has been preserved by God in such a manner as to manifest his special care for his Church."[435]

## Collapse

God will destroy civilizations that curse Him, grind them to dust and blow them away on the wind (Psalm 2:12). But God has made plain that He will not allow anything to destroy His Church, a Church that is in the world, yet not of the world (John 15:19). Thus for the sake of the Church the world is safe until the end of time, and does not need salvation. This good news should encourage environmental realism, not complacency.

Like the uninterrupted state of war and nervous tension Orwell described in *Nineteen Eighty-Four*, endless environmental crises are just as contrived. American people enjoy cleaner water and air, a greater variety and abundance of healthy foods than any time in past centuries. Not surprisingly, therefore, they live longer and healthier lives than ever before. Yet now, more than ever before, invisible radiations and invisible chemicals such as methane or carbon dioxide terrify them. Despite the encouraging facts about the environment, most people *believe* the end is near, and to politicians that is what matters.

After the most massive environmental crisis of all time—Noah's flood—God declared in Genesis 8:22: "While the earth remains, seedtime and harvest, cold and heat, winter and summer, and day and night shall not cease." God promised that despite seasonal climate variability vegetation would never again utterly fail. Such merciful truths should help ameliorate the overwhelming cries of impending doom coming from various authorities and a sympathetic media.

These inconvenient truths leave environmentalists unmoved. If America is not undergoing environmental collapse, they say it is because other people are paying the environmental costs of American excesses. However, the facts get in the way again, for except in places like North Korea, the overall worldwide trend is also improving, spectacularly in places where individual freedom is blossoming.[436]

Well then, with all avenues blocked by facts, the environmentalist constrained by Green principles shows typical human creativeness and argues that though we do not actually see global collapse, it is bound to come because we do not obey the Green Dragon. Paul Ehrlich summarizes with typical hyperbole: "To hell with future generations, and to hell with our fellow human beings today! We'll fly high now—hopefully they'll pay later."[437] Were the argument not so circular, it would be hyperbolic.

No one can deny that human actions have often been shameful towards the rest of the material creation. The flood is a point in case—it was entirely the fault of man. It is not especially pleasant to meditate on the works of man over the past century or so. Wars and forced famines and all manner of injustice abound, with not only human victims. Bison corpses once littered the American plains from horizon to horizon, often killed for little more than their tongues and the pleasure of destruction. In the twentieth century alone several horrific wars plus the work of communist countries splattered onto the ground the blood of hundreds of millions of beautiful humans, sheep for the slaughter. Environmentalists ushered in the first truly Green genocide in the form of technology withdrawal by refusal of Western powers to allow Africans liberty to use DDT. Over 50 million victims have so far paid the ultimate price for slight or no environmental benefit. Their deaths via malaria were preventable. In a sense our technology has finally caught up with our wicked hearts. We have in our power the ability to destroy ourselves, and countless other creatures, many times over.

For decades now environmental doomsayers have been predicting our imminent demise as a species, even specifying the means and the dates. None has been more spectacularly wrong than Paul Ehrlich. About the only prediction where he has been right on is that world population has grown, as he and his wife demonstrate in *Population Explosion*. Any other business would dismiss him as a cracked charlatan; to environmentalists he is a prophet. Unfortunately, since powerful figures in the United States government, perhaps even the 44$^{th}$ President, are disciples of Ehrlich, it is not wise to ignore what he says. In a recent book, *One with Nineveh*, written with his wife, Ehrlich reprises the role of Old Testament prophet. He declares that overweening human pride inevitably leads to degradation of the environment and so to the eventual destruction of humanity and the planet. As evidence for his hypothesis he cites the hubris of ancient Nineveh as a sin found in modern countries, especially the United States.

A Christian can easily agree with Ehrlich that the pomp and pride of modern nation-states is disgraceful, though it is nothing new. Arrogant and secure, we simply ignore God or drown Him out with mindless entertainments and inch-deep religiosity. Pride, security and rebellion: for these, and more, God destroyed Nineveh. There is little hope unless humans repent. But one does not have to agree with environmentalism to recognize the presence of sin in the heart of man. We recognize the sin problem irrespective of environmental or political sensibilities. And environmentalism offers no viable solution. Salvation and renovation are through Jesus, not the dead-end Green Dragon.

Ehrlich purposefully misrepresents, or else misunderstands, the nature of both true preaching and true repentance, both of which have nothing to do with ecocentrism. His Ninevite comparison is more obviously ridiculous when one considers the true hubris of an organic Earth Mother type pretending to be a prophet of God. In the time of Nineveh the penalty for false prophets was death by stoning.

Rather than offering a God-centered alternative—instead of repentance—Ehrlich offers the usual measures "that might allow humanity in general, and the world's sole remaining superpower in particular, to alter course and work toward achieving a sustainable world. Our globalizing civilization urgently needs to explore ways of reorganizing societies."[438] The sustainable world he envisages features command-and-control socialist economics and politics, forced cuts of about 67 percent in global human population, and massive declines in standards of living for those lucky few who survive the transition to the Green workers paradise.

## Nineveh

God spared Nineveh when it repented in response to preaching by Jonah. Jonah preached that immortal souls hung on the edge of hell, "dwelt carelessly," in the language of the Bible. There is no evidence that repentance had anything to do with adopting environmentalist principles. But the story does not end there. Nineveh received much attention from God. In Jonah's era, some 800 or so years before

Christ, it was a great sprawling city and metropolis of a cruel Assyrian civilization. It had a diameter so large it needed a whole day to walk through. The world's oldest surviving aqueduct displays in picture the astounding technological prowess of ancient culture. Sennacherib (c. 720 BC) ordered its construction to bring freshwater from the mountains into Nineveh.

Yet Ninevite glory, like the desert water, sank into the sands and melted away. But God—the first cause—did not destroy Nineveh because of environmental shortsightedness, nor will we face destruction if we fail to heed preaching by Ehrlich and his ilk. God tells us that He turned it into a desert desolation, using various second causes, mainly the ascendant Babylonian empire, to inflict His wrath against a nation that dared harm His holy people (Zephaniah 2:13–14). Rather than a warning to God's people, as environmentalists would have it, God uses the example of Nineveh to comfort His people. Contrary to Ehrlich's claim, the Ninevites did not first turn their land into a desert, and then collapse as a civilization. Scripture plainly teaches that collapse came first, the population declined, and Nineveh finally turned into a desert wilderness when deserted by humans. Finally, the wild animals returned.

Jared Diamond is a University of California geography professor whose writing gives every suggestion of his being Ehrlich's soul mate for all the torrid admiration each expresses for the other. In his bestseller, *Collapse: How Societies Choose to Fail or Succeed*, Diamond argues that many civilizations went extinct because they felled trees, did not practice safe sex, and were environmentally insensitive. Just like Ehrlich's Ninevites.

No one denies that it is possible for societies to collapse due to environmental disaster. Such was the case in Noah's time. Diamond suggests collapse as a rule for societies that do not pay sufficient homage to the Green Dragon. He discounts data that show how political, economic, and cultural forces influence the size and growth of populations discussed, for instance, in Thomas Sowell's *Conquests and Cultures* and *Race and Culture*. Diamond cannot help himself and, like Ehrlich, has an involuntary impulse to employ all his considerable talent to blame Western Civilization, in particular the USA, for being insufficiently sensitive to Green central organizing principles.

The pseudonymous pundit Spengler reviews the book and notes how Diamond is seemingly oblivious to various markers and scientific studies of environmental improvement. The gist of the book, writes Spengler, is to warn "the inveterate consumerists of the United States, who presumably also face extinction should they fail to erect legal barriers to suburban sprawl." Spengler continues, "Ideological reflex is too mild a word for this sort of thinking; perhaps the term 'cramp' would do better. Given that America returns land to the wilderness each year, the danger to American survival from deforestation must be on par with the risks of being hit by a large asteroid. The world is not breeding too fast—birthrates are everywhere falling—and the industrial countries (except for the Anglo-Saxons) fail to reproduce at all."[439]

## Diversity

To the theory of intrinsic value, which requires little or no environmental change, the Green Dragon adds the idea of diversity as another central organizing principle for a new world order. According to Jeremy Rifkin, "Every species must be preserved simply because it has an inherent and inalienable right to life by virtue of its existence."[440] If one raises the reality of increased environmental quality and more forests, for instance, as an argument falsifying the claims of radical environmentalism, Greens counter that there is less diversity, so we must still do as they say. Says E. O. Wilson, "The sin our descendants are least likely to forgive us is the loss of biological diversity."[441] It seems almost unnecessary to point out the inherent illogic of an environmentalist ethos that insists, on the one hand, that humans are mere animals, but on the other insists that humans be the only species to beat its breast for not protecting other creatures at all costs.

Like a knight riding out to search for some nemesis to slay, the quest for diversity is rather nebulous. Perhaps it is the quest for increased numbers of known species. Yet nowhere is it written that all creatures should exist forever. Some of them are downright unpleasant, and we would be well rid of them. Environmentalists demur; they are so committed as to advance their environmentalist doctrine and practice to the status of a universal credo. Al Gore defends this invented imperative by insisting that God highly values biodiversity, else why would He insist that Noah take two of every animal into the ark?[442]

This superficial reading of the Bible is unfortunately common also among professing Christian Greens. Now it is true that God is liberal in His creation and in all His dispensations to His creatures (Psalm 147:9); you cannot look at the creation without recognizing God's maintenance of a riot of glorious diversity. Our fingertips are home to an entire invisible bioculture, one incidentally devastated each time we wash our hands.

But God's creative and comforting liberality, though like rich oil dripping down from an overflowing cup, is not indiscriminate. It is subject always to His glory. His creatures are always at His disposal; He creates them and He destroys them. When Job came to realize this, he rejoiced in God when all other pleasure was gone, and proclaimed: "Though He slay me, yet will I trust Him" (Job 13:15).

With the global flood God commanded Noah to take beasts into the ark, with the reason given in Genesis 7:3: "to keep the species alive on the face of all the earth." So preservation of various species was clearly part of His equation. But when we examine further it becomes clear that biodiversity was a tiny piece of the equation. Only fowls, cattle, and creeping creatures he took into the ark. Though this encompassed much, it left a host of other creatures to fend for themselves. One can infer that, except possibly for food purposes, Noah took limited plant life on board. And whereas God miraculously brought the beasts to Noah, a task he could not accomplish without divine aid, God did not bring to him a plant nursery. The flood thus represents a likely massive extinction event (2 Peter 3:6), mostly in the floral kingdom, and an overall dramatic drop in species diversity.

Biodiversity is thus not an end to itself, nor is it clear that it is desirable in every instance. At best one might call it a second-order good, even if not always and everywhere a good. As the explosion of languages at Babel makes plain, diversity can be a curse. Although this example speaks directly to the historical effect of social diversity, one might extrapolate to other areas. Social diversity, even of something like language, all else remaining constant, resulted in dissipation of the unity of mankind, and mutual confusion and distrust. Diversity is not an obvious contributor to social stability any more than biodiversity is a rule to create stability in an ecosystem.

Of most creatures God allowed a single breeding pair onto the ark, but of the clean beasts he brought seven pairs. It would seem naïve and unwise to leave the future of species in the paws and claws of a single breeding pair and, what is worse, to put them in the hands of Noah. When God drowned the Earth He not only devastated the diversity of the gene pool, but in leaving just a single breeding pair to replenish the devastation (Genesis 7:2) He recreated a world populated entirely by endangered species, a fact whose implications Gore ignores. This left no wiggle room for a species under pressure; one male and one female represents a species technically extinct.

Then we see how Noah, with a view to the proper worship of God, and complete insensitivity to the plight of biodiversity, exited the ark post flood and arrogantly slaughtered some of the endangered animals. He cut their throats, offered them as a pleasing sacrifice to God (Genesis 8:20), and ate their flesh. From the environmentalist perspective this is irresponsible, unconscionable madness.

People influenced, or trapped, by the mental bonds of the Green Dragon could easily conclude Noah was temporarily insane. They might point to his later drunkenness as evidence of a weak and unstable mind, or his meat-eating habit as evidence of a dangerous addiction that would spill over into mental health issues. Professor Peter Singer quotes with approval the following from Albert Schweitzer: "It is good to maintain and cherish life; it is *evil* to destroy and to check life."[443] Nonsense! God creates, and God destroys, and God is good, not evil. The professing Christian Green, who believes God's Word is the only sure guide to truth, must grapple with still another problem—Noah's sacrifice pleased God.

Humans, we can agree, are not divine, in general, and so they may not at a whim exercise the divine prerogative to end life. But human limits add not one ounce of weight to Singer's hypothesis. Precisely what part of the divine command to "kill, and eat"[444] is so hard to understand? In Singer's perfect world it is acceptable to kill the weakest humans[445] but not permissible to eat animals with a clear conscience.[446] According to Singer it is, however, reasonable to have sex with animals.[447] Apparently, if we are to take environmentalism seriously, the command to "love your neighbor as yourself" (Matthew 22:39) reaches beyond humans.

Scripture teaches that destroying animal life, tantamount to reducing biodiversity, is often permissible. Nevertheless, we are not to kill or destroy indiscriminately. Biblical evidence suggests strongly that maintenance of biodiversity is desirable only insofar as it benefits man and brings glory to God. Singer thinks that Christianity encourages indiscriminate slaughter, writing, "Nonhuman animals, on the other hand, were believed to have been placed by God under man's dominion, as recorded in the Bible (Genesis 1:28 and 9:1–3). Hence humans could kill non-human animals as they pleased, as long as the animals were not the property of another."[448] Not so, Mr. Singer. Singer and others would disallow meat eating on invented moral or religious grounds, but Christians can and do kill animals and eat their meat if judged beneficial. Just like Jesus did. If, for whatever reasons, someone chooses a strictly vegetarian lifestyle, that is fine. But that preference should hardly become a moral crusade to impose on others. How dare one oppose God by calling it immoral to eat meats that God has declared clean?

We kill living creatures for other reasons too, not only for food and hide. Concern for species loss is a valid consideration of whether to kill, or stay, not least because living creatures are part of our endowment. But more important considerations always balance such concern. If some creature's existence harms human beings in ways that outweigh benefits, then it is lawful to destroy it. Ignorance makes the decision difficult at times, because we never know all potential harms and benefits. For instance, destruction of sugarcane because its product damages our teeth is clearly not the same as destroying the last bacterium of bubonic plague.

God encouraged the beasts of Eden to be fruitful, infusing them with a great ability to reproduce–their fecundity. Such fecundity ensures that cattle, for instance, are highly valued. Their value is not static, and certainly varies in time. Humans, for example farmers, will sometimes risk their own precious lives to save animals trapped in a burning barn. But they will not take this risk based on some Green imperative. They do so because they have placed economic worth on their animals and wish to sell them to others, who will presumably barbecue and eat them. Until then the cow is better alive than dead. So the farmer resists the panting flames, which hunger to consume but pay with no coin of this world.

In deciding death and life we need unbiased judgments based on the best calculations of the value of a living creature, and therefore the value we might lose (or gain) by its death. The formula is never so simple as Meister Singer's "as they pleased." My personal preferences might not match those of my neighbor. He might use ants to brew alcohol, as they do in Korea, and so be comfortable with some ant-inflicted damage to his property. To him the benefits outweigh the costs. If the law protects my private property rights, and provided my neighbor keeps his ants away from my house, who am I to deny his pleasure?

Or, I may destroy ant or rat colonies that infest my house. In every case one should first always weigh costs versus benefits. This may be done on a personal level but, depending on the extent of risks and benefits, can extend beyond the individual, family, or church, even to the state. Olden day mining companies would take canaries into the pit as a sort of chemical detector. Many canaries perished this way, but their deaths saved human lives. No doubt some miners had regret at the death of the canary, but one can be sure they had no doubt about the righteousness of the fatal choice. If the canary were a rare species, perhaps it is better to prefer an alternative sacrifice. A government holding the last sample of a killer virus might choose to make it extinct, or choose to keep it for research. Diversity should be one of the least considerations in the decision-making process. Clearly it is not always easy to make the correct decision, but we humans have the right, even the duty, to decide, to pull the trigger.

In the end, species that serve human purposes and desires more effectively survive while those that do not tend to decline or disappear. Consider sperm whales, whose spermaceti was an important nineteenth-century source for public lighting and machine lubrication. Thankfully, from the whales' perspective, humans did not really want spermaceti. They wanted light to read by, and efficient greasing materials. Spermaceti only became a resource when some brilliant human turned it into a useful commodity. Modern environmentalists, those macabre Malthusians, fail to see that it is not the resource that people want but the function the resource performs. An idea, not spermaceti, is the crucial input. Lard, plant oils, and then petroleum supplanted use of spermaceti and saved the whales from extinction. Was it good to save them from extinction? I think so, for the least reason that we humans today draw great pleasure from the aesthetic aspects of whale watching.

Environmentalists who fret over whaling need to realize that destruction of one species can enormously benefit many others. Sperm whales are prodigious consumers and eat around three percent of their body mass each day, a significant quantity for a creature that can weigh up to 60 tons. The total annual consumption of prey by sperm whales worldwide is about 100 million tons—a figure greater than the total consumption of marine animals by humans each year.[449] It is not obvious that their extinction would be to the injury of other creatures, though humans might be sad to see them go.

One might argue, as environmentalists have about humans, that voracious seafood consumption by whales has upset the *delicate* ecological balance of ocean life. Ah, you may think the ocean natural, but it is not. Shrimp may think the water caress feels sweet, but it is not nature. Sperm whales plow the seas from pole to pole and have made every molecule of the ocean inauthentic. It is a silly argument, but environmentalists make the same silly argument an art form. They use a similar logic to extrapolate that extinction of our species would return the cosmos to a balanced natural state.

## Beautiful people

When he founded *Greenpeace* in 1971, with several friends, ecologist Patrick Moore was still idealistic about the possibilities of the Green movement. As he gradually became aware of the misanthropic currents of modern environmentalism he quit in disgust. He says the movement is no longer concerned with the environment but is "a political activist movement."[450] In another place he writes, "In the name of 'deep ecology' many environmentalists … perpetuate the belief that all human activity is negative whereas the rest of nature is good…. This aspect of environmental extremism leads to disdain and disrespect for fellow humans and the belief that it would be good if a disease such as AIDS were to wipe out most of the population."[451]

Of all the fixations of the Green movement, none is so dangerous as the concern to limit human population. Hardly an environmental story goes by without getting into a frenzied froth over the population. This focus is integral to the environmentalist movement, towards which the professing Christian environmentalist shows much ambiguity. Baptist minister Oliver Thomas wrote a *USA Today* opinion piece in which he granted the usual. He asked, gently, "In the interest of preserving our planet and our species, shouldn't religious organizations be encouraging smaller families?"[452] He begged preachers to lay on some tough love, to teach that having more than two children "is selfish. Dare we say sinful?"[453]

In his attractive and folksy *Serve God, Save the Planet*, Christian physician Matthew Sleeth repeats the Green mantra in a failed try at humor: "Mothers who have had two previous, uneventful deliveries should quit while they are ahead."[454] The message, to his readers, said with a wink and a smile, Bible verses sprinkled liberally like Christian candy, is that smaller families are not only better for the environment, but also for your health. They are not calling for abortions, but seem fixated on the number *two*.

Environmentalists are now particularly panic-stricken because every expectation is that "the global population will grow 50 percent to 100 percent this century, and per capita consumption of energy and materials will likely increase with wealth."[455] Growth is expected only in non-European races. Europe and the United States are growing only because of immigration, the native-born population failing to breed presumably because they are too busy maximizing their pleasure; they are saving the planet for other people's children.

The perennial problem: too many people with mouths open like baby birds, selfishly demanding a refill. Solution: deliberately cut the numbers of the useless eaters, hamstring the economy, or cut access to electrical power, which will amount to the same thing. Environmentalists believe, with little but passion and blind faith to recommend it, that "consumption of energy, materials, and chemicals accompanying economic growth, technological change and free trade—the mainstays of globalization—degrade human and environmental well-being."[456] Not surprisingly then, at the December 2009 United Nations Climate Change Conference in Copenhagen, population control was raised by Zhao Baige, a member of the Chinese government delegation, and vice-minister of National Population and Family Planning Commission of China (NPFPC). She said that reducing population is an important key to saving the planet from global warming. According to Zhao since the institution of China's one-child policy in 1979 the communist nation has reduced the number of births by 400 million, resulting in 18 million fewer tons of $CO_2$ emissions per year.[457] Aborting humans, she suggests, is the most cost-effective way to save the planet, a much bigger bang for the buck than technological patches or tree planting.

## The Holdren Bomb

Jonathon Porritt, adviser to British Premier Gordon Brown, says, "Each person in Britain has far more impact on the environment than those in developing countries so cutting our population is one way to reduce that impact."[458] He recommends that the country can suffer just 50 percent of the current population, so needs a cut of around 30 millions. Exactly when, or how, the British government hopes to achieve the cuts is unclear.

Al Gore thinks that salvation requires that we "redefine our relationship to the environment, stabilize human population, and use every possible means to bring the earth back into balance."[459] Every possible means? He is not super clear about the details of these means, but abortion and reeducation play a large role. The Sci-fi Channel even plays its part in increasing the volume via dramatic productions like *Countdown to Doomsday*, where narrator Matt Lauer (of *The Today Show*) says, "The stark reality is that there are too many of us. And we consume too much.... The solutions are not a secret: control population, recycle, reduce consumption."[460]

Someone more transparent is President Obama's newly handpicked director of the *White House Office of Science and Technology Policy*, the "Science Czar," John Holdren. Holdren is a Professor of Environmental Policy at Harvard University and an acolyte of Paul and Anne Ehrlich, with whom he has produced many publications that parrot the environmentalist line. In their 1977 book *Ecoscience*, Holdren and the Ehrlichs proclaim their neo-Malthusian and Marxist goals of "population limitation and redistribution of wealth."[461] In another article, written two decades later, Holdren and co-authors say that "[e]xcessive population growth," is "a condition now prevailing almost everywhere."[462] He believes that government pressure is warranted to force a limit of two children per family, writing, "If some individuals contribute to general social deterioration by overproducing children, and if the need is compelling, they can be required by law to exercise reproductive responsibility."[463] The number 2.1 is magical for it represents the zero growth scenario. They consider anything above this irresponsible and selfish.

Population is a monomaniacal obsession for men like Ehrlich, who can hardly keep himself from referring to it on most pages of his voluminous writings. To his credit, none are so consistently honest or realistic as he. He does not flinch at the implications of his philosophy, as do the professing Christian Greens. Says he, "Population growth is a nearly ubiquitous, but all too often ignored, driver of environmental and social problems."[464]

Like Holdren,[465] several decades have done little to moderate Ehrlich's views on population. In the late 1960s Ehrlich wrote, "We must rapidly bring the world population under control, reducing the growth rate to zero or making it go negative."[466] In 1990, the Green tide continued to rise, but still seemed far from the shore. This was the year that Christopher Manes wrote, regarding environmentalist priorities: "most important, the reduction of the human population to an ecologically sustainable level."[467] One wonders if Ehrlich in his wildest dreams ever imagined capturing the ear, via Holdren and others, of the President of the United States of America in 2009. Maybe it surprises him it took so long.

Though he admits it is hard to calculate the population optimum, he is willing to give it a shot, arriving at a maximum number of 2 billion.[468] At our present numbers, this means that 4 billion need to go. According to Ehrlich and others, we have a full-blown planetary emergency on our hands, with the planet on life support, and declining each day. Compared to the terrible 1960s when he wrote *Population Bomb*, matters have only become more critical.

It is intriguing to note that Ehrlich has kindly rounded up his estimates; in 1968 he felt that 500 million would be best.[469] So has Holdren; in 1973 he felt a number in the mid-200 million[470] range would be best, but decades later gave his blessing to an upper limit at 10 billion.[471] This change of thinking reflects an interesting fact about the population question: it is hard, even for types like Ehrlich and Holdren, who have spent their entire adult life obsessing about these things, to come up with reasonable estimates. The simple reason is that their numbers are pure speculation, thus worthless and arbitrary. They really have no idea.

## Multiply

As their dependents multiplied Abram and Lot found that the land could not accommodate their lifestyle. Lot and Abram found that more people creates difficulties, as most new parents appreciate (Ecclesiastes 5:11). But more also create the potential to rise to undreamed heights. There are certainly moments when the sacrifice needed to raise a child becomes a heavy burden to parents. But parents are usually appreciative of the marvelous blessings of a new baby. It must have been grievous to separate as they did. Genesis 13:6, explains: "Now the land was not able to support them, that they might dwell together, for their possessions were so great that they could not dwell together."

And so they separated into two bands. Lot chose for himself land that resembled paradise on Earth, "like the garden of the LORD" (Genesis 13:10). Lot had it made. There was no trouble with carrying capacity. But what followed involved much grief, with loss of gold and life and collapse of Sodomite civilization. This should have been precisely the situation that Jared Diamond might have chosen to explore in *Collapse*. Ehrlich could have written a book called *One with Sodom*, but his political savvy possibly overruled his environmentalist sensibilities. The data for the area of Sodom are better than any other Ehrlich or Diamond considered. Did Lot and the Sodomites degenerate into poverty and extinction because of that fateful separation? Was it really brought on by environmental challenges?

In the shadow of the pressures of life God comforted Abram after the separation from Lot. He reminded Abram of the promise that he would have much land as his own private property. And then, perhaps most remarkably, God said (Genesis 13:16), "And I will make your descendants as the dust of the earth; so that if a man could number the dust of the earth, then your descendants also could be numbered."

The message to Abram was to multiply. Such multiplication, especially given his circumstances, would be an act of confidence in the future. People consumed with fear for the future do not have babies. Environmentalists therefore obsess about too many people. Abram saw glory and sought fruitfulness. One's opinion of children, intolerable burden or immortal blessing, depends on the values of whoever makes the judgment. As David Goldman has written, "Animals breed by instinct, but faith in the future is a precondition for the reproduction of human society. Wounded animals crawl into a hole and die; humiliated cultures turn sterile and pass out of memory."[472]

The reason environmentalists are so mistaken is that, thinking we are bugs in a shrinking Petri dish, they fail to appreciate the potential of human beings made in the image of God. Julian Simon, of the University of Maryland, argued, "It is scientifically wrong, outrageously wrong—to say that 'science shows' there is overpopulation (or underpopulation) in any given place at any given time. Science can only reveal the likely effects of various population levels and policies."[473]

## Invention of a Concept

That word, "sustainable," echoes everywhere. It is the trendiest theme at all the fashionable cocktail parties where Hollywood types, politicians, corporate executives, and college presidents natter about it in reverent and knowing tones. What does it mean? Naturally, given the broadening character of the environmentalist movement, one expects that not everyone means the same thing by sustainability or its synonym, sustainable development. A *World Bank* publication, *Defining and Measuring Sustainability*, provides a nice summary of possible enigma variations. Some think it means a prudent and wise use of resources. Peter Wood explains the dominant definition: "A condition that arises when capitalism and hierarchy are abolished; individuals are made to see themselves as 'citizens of the world'; and a new order materializes on the basis of eco-friendliness, social justice, and new forms of economic distribution."[474] Sustainability is thus a socialist political program thinly disguised in a Green façade.

Sustainability entered the Green lexicon in 1987 with publication of *Our Common Future*, by the United Nations World Commission on Environment and Development, also known as the Brundtland Commission (masterminded and coordinated by Maurice Strong, introduced in Chapter 2). It defined sustainable development as "development that meets the needs of the present without compromising the ability of future generations to meet their own needs."[475]

This definition sounds reasonable at first. None of us wishes to live in a pigsty of our own wastes, and we want to bequeath a better and healthier world to our children than we ourselves entertained. And an improving world has, in fact, been the general experience for people in Western Civilization for much of the past three centuries. On balance, free men consistently produce more than they consume, and thus contribute to increasing prosperity and environmental quality. But one must not fall into the trap of thinking the words from the Brundtland Commission actually mean what they say. The problem is that words do not translate into unequivocal policies. People who have already shown notable skills in semantic gymnastics make words mean what they like to serve an inherently dishonest central organizing principle. In *The Meaning of Sustainability*, Holdren, Daily, and Ehrlich make clear, for instance, that development as an environmentalist code word "should by no means be considered synonymous with economic growth,"[476] but could have more to do with progressive politics. Green phrases often mean far more, or less, than the words say. Instead of creating jobs and economic growth, the Executive Director of *Greenpeace* explained, "We will definitely have to move to a different concept of growth."[477] In such Orwellian terms one sees that development does not mean growth or jobs, and growth means downsizing and layoffs.

Whatever the words mean we still have a major problem with the official Brundtland definition, for it pretends to know what the future will bring unless human development is reduced to zero, or reversed. What the environmentalist fails to appreciate is that human development, particularly through application of Christian principles, is indispensable to the future of the planet. "Stop the world, I want to get off," was the less arrogant cry of another era. This motto at least makes the decision to exit a very personal one, and recognizes that in one's absence the world could continue in its course. Sustainability gurus have a new mantra, simply, "Stop the world." In this view human development that matters must stop, freeze in time like a mosquito trapped in amber.

"If we want to save the planet," writes Robert Fowler in summary of the Green agenda, "we must turn to an economics that will speed the arrival of sustainability, not growth or expansion."[478] In *The Greening of Protestant Thought* he notes how, "In Protestant environmentalist thought today there is considerable hostility toward capitalism. ... Usually the word 'capitalism' is not mentioned. The softer phrase 'market economy' is the term of preference, but the message is the same: capitalism causes, encourages, and justifies using and abusing the environment for human purposes."[479] All the mainline Protestant denominations have issued statements that echo the secular environmentalist attacks on the economic system of free and uncoerced exchange of goods and services. With the humiliating collapse of the Soviet Union almost two decades ago, it appears that socialist liberation theologians have found in the sustainability craze a more than suitable substitute for their ambitions.

Sustainability matured at the second Earth Summit, in Rio in 1992. Three important documents came out of this meeting: The Convention on Biological Diversity; The Framework Convention on Climate Change; and Agenda 21, "an encyclopedic document ... setting out the basic framework for a comprehensive program of actions."[480] A major problem with these documents is that they seek to override laws of sovereign nations and bypass the consent of the governed by putting policy making under the control of individuals who agree with the Green agenda. It becomes clear, as wrote the President of the Czech Republic, that "[t]he themes in the contemporary dispute (or perhaps clash) are clearly about human freedom—not about the environment."[481]

Sustainable development, defined by these documents, wants a system of government in which the input of the little people is neither sought nor required. Maurice Strong is particularly enthused by the potential of Agenda 21, his action plan to "reinstate in our lives the ethic of love and respect for the Earth that traditional peoples have retained as central to their value systems and revitalize the values common to all principal religious and philosophical traditions."[482]

These documents "provide the basis for a new global partnership for sustainable development, enabling the world community to make the transition to a secure, sustainable and equitable future."[483] One of the recommendations was the creation of national strategies for sustainable development. In July 1993 President Clinton complied with this recommendation by establishing, via executive fiat, the President's Council on Sustainable Development (PCSD)[484] to help cities and towns develop their own Local Agenda 21 for a sustainable America. There currently exist several hundred national councils on sustainable development, all coordinated by Maurice Strong's organization, Earth Council.

"Sustainability" appears innocuous at first, but it roars into focus as a great Green Dragon that insists on urgent, and intrinsically violent, changes in social institutions, economic structure, and finally, human consciousness. As Al Gore has written, "our relationship to the earth may never be healed until we are willing to stop denying the destructive nature of the current pattern."[485] We could agree with the letter, but not the spirit, of these words. Living to eat, to consume, or living to save the Earth, is all vanity. If we do not live for God, which means to be taken up by Christ and His law-word, we cannot truly claim to love Him (John 8:51; 14:15). Now if our relationship with God is hostile, all other relationships are hopeless; our only hope of healing is in Christ, otherwise God offers no hope for, says He, "all they that hate me love death" (Proverbs 8:36).

Indeed, what Christian does not grieve at national indifference, even hostility, to the King of kings? Lovers of God would rejoice if nations would cease from murder of innocents, lay down their tools, and honor the Lord's Day. As God says, "turn away your foot from the Sabbath, from doing your pleasure on My holy day, and call the Sabbath a delight" (Isaiah 58:13–14). Our rulers hinder the progress; they follow evil counsels and resemble parasites and oppressors more than nursing fathers and mothers (Isaiah 49:23). We live in a day when every man does what is right in his own eyes (Judges 17:6). Where then is the healing? And so, Gore is technically correct.

But what he is saying here is more than it first appears, and altogether opposite to the Christian view. He recognizes the spiritual death rattle in Western Civilization, and offers baptized environmentalism as the antidote. Just as Heidegger preached "authentic being" as the path towards an expanded or elevated consciousness, Gore preaches "sustainability" as the means to personal healing. He may adopt the language of the Bible, but his principle speaks nothing to the well-lived life dedicated to loving Christ. We can agree that pollution for the mere sake of expansion of power and influence over other men is evil. But this too is not what he is saying. Indeed, he sometimes uses China as an example for materialist westerners to emulate when China is itself an environmental basket case. What Gore is criticizing here is the free and uncoerced exchange of goods and services and our current pattern of having babies without government permission. Sustainability as a concept, and as a political project, extends far beyond humble wishes for cleaner air and water. Environmentalism has become the new ugly face of international socialism, a blunt tool of those Green with envy.

Never mind the facts, which show that never in the history of mankind has an (imperfect) economic system such as capitalism done so much, with so little, for so many. Like their secular contemporaries, professing Christian environmentalists believe a change in human values and consciousness alone can save the planet. They believe the psychological change will lead to institutions and lifestyles that create peace and harmony in human lives and the planet. The only difference, it seems, is that professing Christians add Jesus as another ingredient to their peace pie. They will gag on it, for Jesus is an all-or-nothing kind of God. Environmentalist principles will never yield peace and planetary harmony. The Green Dragon knows this. As it is written in Isaiah 57:21, "'There is no peace,' says my God, 'for the wicked.'"

## Conclusion

Biblical data do not sustain radical environmentalist goals for society, namely stasis or sustainability. Sustainability ignores God's hand in history and pretends that society is amoral and static. But the Bible teaches that God has an interest in the behavior of humans, and also that "It is a fearful thing to fall into the hands of the living God" (Hebrews 10:31). We may be grateful that God is a merciful Sovereign, and slow to anger, for if He curses a society, no amount of personal or central planning can save it from disaster. History is therefore not a matter of fate, conservation, or man's abilities, but of the decree of God.

At the same time, God commands men to take dominion in the name of Christ, to fill the Earth and multiply. This is not possible without a wholehearted embrace of Jesus Christ, on a global scale. While we live in a day of great wickedness and of small things, spiritually speaking, that is no excuse to claim that slavery is necessary to prevent the potential sinful abuse of power. After all, God commanded men to redeem the world even after the fall. Christians believe that people are already in prison; changing cells will not change that fact. Rather than seeking to save the planet on terms dictated by the Green Dragon, Christians ought to be preaching the only message that can save the planet—the gospel of liberty in Jesus Christ.

People who choose to believe that the God of the Bible does not exist are fools (Psalm 14:1). It is therefore a madness to despise and mock God. Reformer John Calvin insisted, "there is no stupidity more brutish than forgetfulness of God."[486] Such a powerful abandonment of truth results in an intense ethical and spiritual dissonance. Even the devils know God, and they are smart enough to tremble at that knowledge (James 2:19). Not only do atheists and agnostics, by their unbelief, call God a liar, but they also deny what is self-evident in nature; they deny empirical data and reason. Still blessed with a semblance of reasonableness, they are able to arrive at a host of logical, but destructive, conclusions when starting from the axiom that God is a phantom of evolutionary dreams. This, much more than materialism, is the reason for the sense of spiritual emptiness and discontent. Materialism is a symptom, not a cause, of a spiritual problem.

In its heart of darkness the Green Dragon believes man thinks too much of himself. Man *qua* man is unacceptable—man must be remade in the image of the Dragon. This low opinion of humanity was mentioned previously, humans being called cancer, plague, virus, and other sweet nothings. If these characterizations are sincere, should it be surprising that there is little mercy for our species? Despite its noble rhetoric about saving the planet, or doing good for the children, more than anything else the doomsday obsession with a culture of death shows the ugly face of the Green Dragon.

Environmentalists warn of a planetary collapse unless humans adopt what Denis Hayes calls a "common consciousness"[487] of reverence for the ecosphere. One might hope that Christian contributions to the discussion on proper human relationships to the rest of creation would swerve towards truth and avoid such pagan nonsense. However, most books written by professing Christians adopt, rather uncritically, the most radical of environmentalist positions and solutions, if not all aspects of the Green philosophy. Whether one reads Ian Bradley's *God Is Green: Ecology for Christians*, Cal DeWitt's more recent *Earth-Wise*, Matthew Sleeth's *Serve God, Save the Planet* or his more recent *The Gospel According to the Earth*, Jonathan Merritt's *Green Like God*, or most else, the argument is invariably the same. Basically, the world is going to environmental hell, it is almost entirely the fault of humans (mainly Americans), and a fundamental part of repentance is, in the words of the *National Council of Churches* (NCC), to "treat creation as sacred." Until we reject the false gospel, says the NCC, "a gospel that proclaims that God cares for the salvation of humans only,"[488] all hope of Green salvation is gone. So it is that they replace true saving gospel repentance with politically correct platitudes.

As further evidence of our sincere repentance Greens urge adoption of so-called sustainable practices, such as dedevelopment of industrial society. As Thomas Derr explains in *First Things*, what Greens mean by this is opposition to economic growth, including in poorer countries. He says that there is a "long history of environmentalism showing a contempt for the aspirations of the poor, and in our time strongly opposed to economic development in the weaker countries lest global environmental harm result."[489]

Second causes for civilizational collapse are surely varied. One that Ehrlich, Diamond, and their ilk discount is a death wish. Europeans will not make babies for love or money, and most other Western countries, or those that have adopted Western mores, trend in the same direction. In *The Empty Cradle* Phillip Longman presents his evidence that modern Western Civilization faces massive demographic changes followed by depopulation on a scale not seen since the Roman Empire collapsed. He discovers, to his dismay, "Faith is increasingly necessary as a motive to have children."[490] And, to be clear, Longman is not talking about faith in the Green Dragon. As we have seen in previous chapters, faith in the Dragon does not lead women to embrace their fertility, but to abort their children to save the planet.

People who for no physical reason fail to breed have lost all hope in the future. Suicide among animals is hardly known, but this curse is a growing affliction among humans. It is not surprising to find such behavior from a people that have absorbed the Green lessons.

The agenda of the environmental movement is to cut population and reverse population growth, whatever it takes to do so.

Unfortunately the Green vision extends beyond abortion of children in the West. After a lull of eight years the Obama administration is now reloading population control programs in poorer countries. For decades, and now with a vengeance, environmental activists have used their influence with the U.S. State Department to activate population-control programs that use aid money to encourage governments of less developed countries (LDC) to abort their children—all in the name of saving the planet. As if the blood of fifty four million murdered babies in America was not enough, we are once again witnessing globalization of infanticide.

Instead of saving the world "for the children" the road to Green hell is paved with the skulls of children. Many politicians in LDCs are fully on board and content to carry out and praise the benefits of population reduction. So long as the flow of blood money keeps on coming they will hold the line. A notable exception is Ugandan President Yoweri Museveni who said, in July of 2008, "I am not one of those worried about the 'population explosion'. This [population] is a great resource."[491] Now if only Uganda were to know great spiritual revival, of the type that turned European barbarians into Christians, there would be every reason to expect emergence of an African miracle. When Western barbarians turned to Christ and adopted Christian principles of liberty and truth, the event fixed the greatness of Western Civilization. In time it will occur again. Once a man looks on Christ with a spiritual eye he goes away quite changed.

In *The Great Global Warming Swindle*[492] documentary, Kenyan economist James Shikwati said, "there's somebody keen to kill the African dream. And the African dream is to develop." The Green Dragon is keen to kill a lot more than dreams.

# CHAPTER SEVEN

# Dominion, Not Death: How Environmentalism Harms the Poor and Biblical Christianity Offers the True Hope for the Future

*In [the ancient pagan world], many saviours were offered by many religions to the attention of men, but the various pagan religions could live together in perfect harmony; when a man became a devotee of one god, he did not have to give up the others. But Christianity would have nothing to do with these "courtly polygamies of the soul"*

J. Gresham Machen[493]

## Snow Angels

There will be snow. Frigid temperatures or extreme winter weather seem to haunt major global warming conferences and events. Laughing researchers describe this as the so-called "Gore Effect." For instance, soon after his arrival at the December 2009 United Nations Climate Change Conference in Copenhagen a blizzard joined in the fun and dumped 4 inches of snow in less than a day. Such cases abound. A snowstorm nixed a Capitol Hill media briefing related to a U.S. Senate climate bill in March 2007. On October 22, 2008 Gore brought a global warming speech to Harvard University plus the coldest temperature ever measured there. Less than a week later, and just as untimely, Big Ben in London enjoyed the first October snowfall since 1922, ironically during the British House of Commons debates on global warming. In January 2009 Gore braved snow and a major ice storm to urge lawmakers in Washington, D.C., to pass laws taxing greenhouse gas emissions.

On March 2, 2009 the biggest snowstorm of winter in the American East disrupted the lives of tens of millions, including a global warming pity-party in Washington. Schools closed in the North through to the Deep South because of roads covered with black ice and snow. Alpine mittens, inches of snow underfoot, and all around snotty noses were the order of the day in Washington among the die-hard environmentalist urchins of Capitol Climate Action. They were there to shut down the coal fired power plant associated with the government complex on the Hill. Instead, they were frozen out.

We know from experience that nowadays media will cite a blistering midsummer afternoon, or a winter's day slightly warmer than usual, as evidence the end is near. But the environmental alarmists keep getting their dates hilariously wrong for their global warming protests, and winter 2009, like winter 2010, was singularly uncooperative. It is hard to argue the ocean is boiling during a winter storm, best to time it for the height of summer. God does indeed have a sense of humor. One large banner read "Stop coal>>Solve Global Warming." Another protester held a sign simply reading, "No coal," in his frigid left hand, green hard hat on his bobble head, and a "Climate Justice" sign clenched in his shivering right hand. This red-nosed Rudolph was no snow angel; he wisely retreated indoors before all the global warming falling from the sky froze his dainty bits. Al Gore strikes again.

## Know Pain, No Gain

All the big money is riding on the Green tsunami politicians wish to amplify. President Obama has promised a revolution for the United States and the world, with a plan that he says will force the price of electricity to skyrocket. He aims for a "new energy economy,"[494] which distills to the simple idea that you pay more, much more, for energy and everything that depends on energy, and use less, much less. Just to be clear, this is not a left-right issue. Senator John McCain, the Republican Party candidate in the presidential campaign, promised much of the same pain.

On January 17, 2008, in an interview with the *San Francisco Chronicle*, Mr. Obama promised to merge the powers of government and favored industries to cause massive rate increases for Americans through an environmentalist plan called cap and trade. Cap and tax or tax and trade seems more appropriate. A proposed $500 million rate increase to affect the Carolinas is an early signal of the incredible pain that politicians could inflict on a global scale. In their regulatory request representatives of Duke Energy, the third largest electricity utility in the USA wrote, "In addition to the significant costs associated with existing state and federal environmental and other regulatory requirements … we are facing expected greenhouse gas reduction requirements in the near future."[495] Of all their challenges, they write, "[o]ne of the most significant … is the cost and technological obstacles of compliance with anticipated climate change legislation."[496] Another contributor to the Duke report writes, "Greenhouse gas regulation is the fulcrum of all the major challenges we face."[497]

Duke is under pressure not least because it is difficult to run a business for long with massive rate hikes. It is unsustainable. The reality is that "[i]mplementation of a new federal carbon policy will significantly increase the cost"[498] of electricity generation. And the victim of bureaucrats flexing their muscles, Duke makes clear, will ultimately be the customer. It is not the federals alone, but minor bureaucrats, such as Los Angeles mayor Antonio Villaraigosa, who also want to express their Green pride by rate hikes. By the end of 2010 Angelinos could be paying up to 28 percent more for electricity, to fund the mayor's new environmental initiatives.[499]

South Africans know something of what Americans may soon face. They have suffered daily indignities brought on by Eskom, the undependable government power utility. In 2008 the South African government called on customers to reduce energy use by 20 percent.[500] The bungling chairman of Eskom, Jacob Maroga, explained that a sudden 32 percent increase in electricity rates will ensure that people comply. Maroga says that it is really for the best;[501] the rate hike plan will simultaneously force consumers to reduce their use of power (good for the planet), raise funds for Eskom (good for the government), and limit rolling blackouts (good for consumers). Actually, only the government and its experts win, for they now have flexibility to redirect financial and energy assets to places they judge most deserving, like the World Cup soccer stadiums.[502] Ecotourists to South Africa no doubt enjoyed a floodlit 2010 World Cup and modern game reserves with all the usual wild animals the country is so famed for. Had they cared to rough it a bit, they could have explored the human game reserves outside the electrified area, where powerless South Africans sat in the dark.

Incompetence and government interference often have similar outcomes. In America the government penalties the Obama plan promises will offer indignities not unlike what the South Africans now suffer. To save the planet it is imperative to make energy so expensive that people can hardly afford to use it. This is why a *Time Magazine* report challenges politicians to do all they can to "raise the price of gas. We already know that higher gas prices discourage driving and reduce greenhouse-gas emissions."[503] In its budget for 2010 the U.S. Environmental Protection Agency (EPA) set a target to cut carbon dioxide emissions in the transportation sector 14 percent from 2005 levels by 2020. If effected, according to Belfer Center for Science and International Affairs at Harvard University, fuel taxes must rise in the USA, to ensure prices at the pump are at least $7 per gallon.[504] At the United Nations conference in Copenhagen in 2009 Tim Wirth, a former US Senator and close colleague of Al Gore, hinted his approval of this strategy, citing the "enormous regulatory authorities at the EPA."[505]

Neither Wirth nor the *Time* reporter appears concerned about the resistance or suffering of the little people. But those poised to deliver the expensive, inefficient, and unreliable government-mandated green energy are bracing themselves for anger at the coming brownouts. Keith Martin, a lawyer at New York-based Chadbourne & Parke LLP, who represents developers of alternative energy projects, says it will not be long before "we could experience ratepayer backlash."[506]

## Blood and Gore

In *The End of Affluence*, written in 1980, Paul Ehrlich stated, "One general prediction can be made with confidence: the cost of feeding yourself and your family will continue to increase. There may be minor fluctuations in food prices, but the overall trend will be up."[507] The reality is that long before Ehrlich wrote the book the real trend was for the quantity, quality, and availability of food to increase and improve. And the relative price of food has been going down, and not only in Western nations. Never mind; it seems that politicians are recklessly determined to enforce this prophecy by denying the free and uncoerced flow of goods and services. In the United States, for instance, government environmental shackles prevent the people from using their own embarrassing overabundance of coal and oil.

Global warming is a dream come true for busybodies who want to micromanage the details of the lives of others. The new American administration gives little evidence that economic or individual liberty is on the agenda. So far Green central organizing principles appear to be an important basis for actions. After the Holdren bomb President Obama recently appointed Harvard's Cass Sunstein to be his "Regulatory Czar." Sunstein's job will be to ensure compliance to whatever Green controls his office and numerous government agencies can dream up. Unsurprisingly, he believes that "animals, species as such, and perhaps even natural objects warrant respect for their own sake, and quite apart from their interactions with human beings."[508] Naturally, he predicts that we will soon[509] have laws that allow animals and plants to sue humans for violating their rights, an idea previously backed by Czar Holdren.[510] Whether most people realize it, the underlying philosophy informing these positions is hardcore environmentalism. Harvard law constitutional scholar Laurence Tribe explains that his support for such legislation is based on a "deep intuition that chimps and dolphins and dogs and cats are infinitely precious—like ourselves."[511] This is happening all over the place. In fact, in March 2010 voters in Switzerland got to decide whether animals can sue.[512]

Is this just what they need? And can only humans be sued? No word on whether the government will allow worms to sue birds, or birds to sue cats, and which party government lawyers will represent. And what about the trees? That President Obama has selected as confidantes several men of such beastly thoughts reveals his own special belief that enlightened specialists hold the key to saving the world from humans. Men like this spread much corruption and injury to society.

Cap and trade has been in place for a while. The global cost was $126 billion in 2008, and the new cap-and-tax scheme will steal much more from your collective wallet. Carbon trading alone is expected to become an annual $2 to $10 trillion[513] black hole. That works out to be about $30,000 for each American, or $120,000 for the average family of four. Of course, the carbon trading is international in scope, so others will pay too, but the general idea is that the USA will bear most of the costs. You do the math. This is Enron on steroids, and predators are salivating at the profits they will cream off; when government binds citizens in Green fetters, it is simply much too great a temptation to avoid joining in the plunder. A 2007 Lehman Brothers media release predicted, "As carbon emissions trading becomes an increasingly important aspect of managing the challenge of climate change, the opportunities for participating in this market are set to grow exponentially."[514]

Global warming's Dr. Strangelove, the space physicist James Hansen, calls the proposed U.S. laws a "counterfeit climate bill known as Waxman-Markey … a monstrous absurdity … Ponzi-like 'cap-and-trade' scheme."[515] Hansen's protests ring just a little hollow in light of the little known fact that he helped draft Lehman Brothers' "Business of Climate Change" reports, which promised investors huge profits through carbon trading. Democratic Party politician Ben Cardin calls cap-and-tax "the most significant revenue-generating proposal of our time."[516] Senator Cardin is too modest—of *all time* is more like it.

If the plans come to fruition carbon will become the world's largest commodity market, bigger than gold, bigger than oil—trafficking purely in hot air. Like Hansen, the famed environmentalist and scientist James Lovelock sees through the mirage. "Most of the 'green' stuff," he says, "is verging on a gigantic scam. Carbon trading, with its huge government subsidies, is just what finance and industry wanted."[517] Were it really in danger, perhaps one might try to justify such extreme costs of tens of trillions to "save the world." But such self-inflicted economic trauma will do nothing of the sort, not least because the rest of humanity will continue to hobble along[518] even if Americans choose to punch themselves out and turn the lights off. At Copenhagen 2009 China and India pulled the plug on the dreams of not a few wannabe carbon billionaires. Both nations signaled that they will not sign any treaty that might reduce their capacity to increase economic production.

## Sweaty Palms

Prospects of increasing their authority, exponentially if possible, are a fatal attraction for too many politicians. Previous efforts to scare up control via environmental warnings achieved little traction on slippery ground, but now the sky is the limit. Two decades ago Tim Wirth said, "We've got to ride the global warming issue. Even if the theory of global warming is wrong, we will be doing the right thing anyway, in terms of economic policy and environmental policy."[519] Since 1998 Wirth has functioned as president of the United Nations Foundation, funded by a $1 billion contribution from Ted Turner to support environmentalist causes, in the name of the United Nations.

Except for the carbon bandits, saving the planet is all pain for no gain. After trillions of dollars extorted from American taxpayers the expected effect on climate will be so small as to be immeasurable; the world will not be saved, after all. Instead insane government policies will ensure a lot of us will be dead. To be sure, though environmentalists such as Hansen and Lovelock see through the cap-and-tax scam, they are on board with the theory of Green apocalypse. It is not that too much money is washing down the Green Dragon's throat, they want more.

Soon after the 2008 American Presidential election the Commissioners of The Church of England, seeing a sure thing, voted to invest about $230 million in Generation Investment Management (GIM), Al Gore's Green investment arm. They are not stupid, they can read the political bones. The managing director of GIM is David Blood, the former CEO of bailed-out Goldman Sachs. Known as "Blood and Gore," or just "The Carbon Empire,"[520] GIM is rapidly evolving into a megalodon-mouthed whale that will vacuum up an ocean of green cash set to hemorrhage from the blows of environmental policy.

While the Church is getting on the environmentalist bandwagon in a big way, especially the anthropogenic global warming scare, a few brave scientists are starting to jump off. It is likely that data that have falsified global climate models are playing a role in turning the tide; observations show a consistent cooling trend at least from 2000 despite large increases of carbon dioxide, and the jury is still out on whether the trend continues at present. Or possibly they have become offended by the transparent abuse of science by oodles of money devoted to showing that increased atmospheric carbon dioxide will overheat the planet—$79 billion[521] in the USA alone in the two decades since 1989. The corruption of science exposed by the Climategate scandal undoubtedly plays a role.[522] In future, this will be cited as the archetypal case of abusive monopolistic science. MIT meteorology professor Richard Lindzen writes, "The fact that the developed world went into hysterics over changes in global mean temperature anomaly of a few tenths of a degree will astound future generations."[523]

Assuming a free press, it is incredible one hardly hears that cooling of the past decade, if it continues at the same rate for one hundred years, will result in a global mean temperature 4°F lower than at present, and then we will be in an ice age. Much as we would like it, scientists just do not know how to predict climate. But the Sun would have to freeze over before one can pry men like Al Gore off the gravy train they have invented. Arthur C. Clarke, author of many best-sellers such as *2001: A Space Odyssey* summed it up on the cover of a book by distinguished atmospheric physicist Fred Singer, a book that laughs at the Green scam. *Hot Talk, Cold Science*, writes Clarke, "dares to point out that 'The Emperor has no clothes.'"[524]

## Christianity as a Greek Cult

Bjørn Lomborg, Julian Simon, and others have exploded the environmental myth that humans, to use Al Gore's rhetoric, "have begun to wage war on the earth itself."[525] Our sinful nature notwithstanding, on balance human beings are creators, not destroyers. Environmentalist accusations hint at another popular myth, namely, that before Christianity and its spawn—modern civilization—primitive agrarian societies were friends of the Earth. But Western Civilization has revolted against the natural order and breeds Earth defilers and Earth rapists. So part of the solution to a human consciousness that is in harmony with Gaia is to live a simpler life that leaves light footprints on the Earth. This explains the urgent calls to embrace poverty and reject growth and economic or industrial development.

Many professing Christians have absorbed much of the environmentalist appraisal of Biblical Christianity. There is, for instance, sympathy for the more radical postures of Næss's deep ecology with its belief that the idea of self as the image of God is harmful to human life quality. Al Gore speaks of lives filled with intense disharmony and psychic pain because we do not hold to environmentalist sensibilities and practices. "The cleavage in the modern world between mind and body, man and nature," says he, "has created a new kind of addiction: I believe that our civilization is, in effect, addicted to the consumption of the earth itself. This addictive relationship distracts us from the pain of what we have lost: a direct experience of our connection to the vividness, vibrancy, and aliveness of the rest of the natural world. The froth and frenzy of industrial civilization mask our deep loneliness for that communion with the world that can lift our spirits and fill our senses with the richness and immediacy of life itself."[526] Charles Birch and John B. Cobb, Jr., in *The Liberation of Life: From the Cell to the Community*, similarly argued that Christianity is excessively intellectual and thus alienates us from ourselves and from the divine. Green theologian Rosemary Ruether echoes the same theme: "Westernized consciousness must heal itself of its split-off divisions that have separated knowledge from wonder, reverence, and love before we can learn how to tell the cosmic story in a way that will rekindle an ethic and spirituality capable of calling us to the tasks of healing and sustaining the earth."[527]

Philosopher John Passmore argued that there exists in Christianity a dualism between God and nature that is the result of Greek Platonic influences.[528] Plato, he says, encouraged us to think of nature as the "other," separate from the mind. Gore, or ironically his ghostwriter, in speaking of Western Civilization, recycles the same story: "The strange absence of emotion, the banal face of evil so often manifested by mass technological assaults on the global environment, is surely a consequence of the belief in an underlying separation of intellect from the physical world. At the root of this belief lies a heretical understanding of humankind's place in the world as old as Plato, as seductive in its mythic appeal as Gnosticism, as compelling as the Cartesian promise of Promethean power—and it has led to tragic results."[529]

In other words Christianity's alleged inability to resist the hypnotic charms of Hellenistic thinking results eventually not only in environmental disaster, but for humans a loss of purpose, rootedness, and emergence of bone-numbing unhappiness. While they invoke Plato, environmentalists invariably stretch back farther to fancied days of Earth-centered religion. Here lies the rationale for the Green demand that common people embrace a pursuit of purposeful poverty. So in avoiding one heresy Greens fly into the jaws of another. Charles Rubin, a critic of environmentalism, summarizes in *The Green Crusade*: "The root of our ills turns out to extend further than … Descartes and Bacon…. It reaches beyond the Platonic foundations for 'anthropocentrism'…. monotheism is implicated in our present crisis, and the root may extend all the way to a wrong turn we took when we abandoned the hunter/gatherer life for settled agriculture."[530]

But to fast-rewind back to Plato: Is Christianity in its modern form just an admixture of Greek heresies? Many so-called Christian environmentalists think so. That is why they reject man's singular place in the cosmos, and argue for the sacredness of nature. Theologian Thomas Berry wrote, "The first condition for achieving this objective [cosmic harmony] is to realize that the universe is a communion of subjects, not a collection of objects. The devastation of the planet can be seen as a direct consequence of a loss of this capacity for human presence to the nonhuman world."[531] Plato's views do result in an unresolved dualism in which God, the Creator, is somehow unable to control swathes of eternal and chaotic space. Christian authors Mark Stoll and Robert Fowler describe the synthesis of Platonic ideals with medieval Christianity, for instance the *One* of Plato, and the one true God of the Bible.[532] Although Plato's pupil Aristotle, from whom Thomas Aquinas learned so much, admitted matter's importance, he was still hopelessly anthropocentric. He held that humans alone have an active intellect and therefore they alone, as summarized by Peter Singer, "belong on the side of God while all other creation stands on the side of matter."[533]

One need not be so extreme to recognize that from the earliest days there have been theologians who sought to defend the truth of Christianity by adopting (sometimes consciously) pagan philosophical positions. Platonic dreams tempted Augustine, to which Aquinas added some Aristotle. Amid the crumble of Rome, with its ensuing chaos and mass human extinctions, Christian philosophers, according to historian Gary North, "incorporated aspects of Greek wisdom, and therefore Greek dualism, into their defenses of the orthodox faith. The result was intellectual schizophrenia—philosophical syncretism. Christian philosophers attempted to combine irreconcilable systems: Greek philosophy and biblical revelation."[534] Thus the world, and Christianity, entered a long period of darkness. Through the medieval malaise mere glimmers of light finally burst into the glory of the emergent Reformation.

There is no connection between Biblical revelation and pagan philosophy. The one comes by inspiration of God, the other from the wisdom of man, or the suggestions of devils.[535] Pagans reject God's revelation of His divine law for human behavior and, sadly, so do many in mainline Christian churches. God has provided in His words unchanging standards and guidelines for even the most specific parts of life. In the words of Jonathan Edwards, "God knows better than us; and whatever God declares is true."[536] Christian reformation beginning in the sixteenth century sliced off the leech-like paganism latched onto the pure name of Christ. And second, it brought a return to the Biblical teaching that divine revelation alone, not tradition, and certainly not unaided human reason, was the source of truth.

Those who stress pagan Greek influences on Christianity[537] forget that Christianity is the divine original, not some cultural invention. Yes, pagan ideas early invaded the Church of Rome and crop up in strands of Protestantism, but this does not necessarily mean they are native to the Bible. We clearly need another Reformation with reaffirmation of a commitment to the full-orbed truth of the Bible.

Biblical Christianity purges paganism. This does not mean, as some imagine, we are ascetics who read the Bible alone and live on top of a pole. Christians live in time, and in this world of space, so it should not be shocking to allow that in God's providence His Word contains genuine flavors from the many nations His chosen people met. The

Bible is written in Hebrew, Aramaic, and Greek, and these languages alone reflect influences from many other cultures. Why should this be perceived as a weakness or argument against the truth of the words of God?

Gamaliel, the teacher of the Apostle Paul, encouraged in his students more than passing familiarity with Greek philosophy. Paul was expert enough to debate and quote Greek works, for instance mocking[538] the Epicureans who, like modern atheistic environmentalists, exalted the material to be the sole source of truth and meaning (Acts 17). Paul quotes with approval a line from Menander (c. 342 BC) the pagan playwright (1 Corinthians 15:33), and in Titus, chapter 1, he satirically quotes the poet Epimenides (c. 600 BC). Paul debated and refuted pagan Greek ideas in the Greek language. It is thus something of a slight, and a joke, for comparatively inexperienced moderns to think a scholar of Paul's stature was a naïf seduced by pagan philosophy. Quite clearly Paul felt it useful, if only as a point of contact, to be thoroughly conversant with pagan ideas, even if he rejected those parts that were against Biblical truth. Where Paul demolished vain philosophy, our modern age is finding itself captured by the same.

## Flesh

During his phase as a professing Christian, Al Gore asked, "Why does it feel faintly heretical to a Christian to suppose that God is in us as human beings? Why do our children believe that the Kingdom of God is *up*, somewhere in the ethereal reaches of space, far removed from this planet? Are we still unconsciously following the direction of Plato's finger, looking for the sacred everywhere except in the real world?"[539]

Gore's solution to all our problems—moral, ethical, spiritual, environmental—is to reach beyond the Greeks, back to the ages of supposedly Earth-focused religion. Platonic philosophy, he argues, "assured that later explanations of the workings of the world would become progressively more abstract."[540] He wants a respiritualization of our low view of the physical. God's Word is not, however, a product of Greek philosophy, so going beyond the Greeks to earlier animists is futile. Going back to Jesus, to His Word, is what we need. That is what the Apostle Paul did. But it is not what Christian Greens do—they call for revisions of historical Christianity. Lynn White, who arguably started the modern mainstream Christian stampede to environmentalism, wrote, "Since the roots of our trouble are so largely religious, the remedy must also be essentially religious, whether we call it that or not."[541]

Despite criticisms that Christians degrade the value of nature,[542] the Bible does not renounce the world this side of the grave. In His Word God everywhere admits the reality and importance of the material. He reveals to mankind the proper perspective, which is neither Platonic nor Aristotelian, neither idealist nor materialist. Matter is indispensable in the redemptive story; we recall that not only did Jesus come in the *flesh* but He rose to glory in a body of flesh. In addition, the resurrection bodies of Christians must be real flesh, though exalted above anything we can now imagine.

Perhaps some Greens imagine that whatever they see in any church is Biblical. But pagan ideas and practices are common. Almost every tourist who has climbed the North Tower of *Notre-Dame de Paris* has taken a photo of gargoyles grinning down at passersby far below; folklore suggests that "gargoyles may be survivals of pagan beliefs the Church permitted to persist besides Christian subjects, incorporated into church decoration for superstitious reasons."[543] Pope Gregory[544] (c. 540) directed priests to use pagan temples but to Christianize their symbols, but as churchmen have experienced around the world for centuries, the strategy of worldly accommodation might succeed in getting pagan bodies into pews but often fails to remove pagan beliefs from their minds. Take for example the growing embrace of vegetarianism not from personal preference but as moral conviction—a conviction promoted by Hinduism and now by environmentalist biological egalitarianism. Because it advocates abstaining "from foods which God created to be received with thanksgiving," which includes meats (Exodus 12:3–8; Deuteronomy 12:15; Acts 10:9–15), it is unbiblical (1 Timothy 4:3).

Says Wendell Berry, "If we think of ourselves as lofty souls entrapped temporarily in lowly bodies in a dispirited, desperate, unlovable world that we must despise for Heaven's sake, then what have we done for this question of significance?"[545] In short, according to Berry, and others, the problem with Western Civilization is that we have cut ourselves off from the physical with all its juice and emotion in favor of the coldness of rational intellect. Insofar as these criticisms challenge unbiblical practices of Christians, they are valid, and useful. The paganism and false pietisms of every branch of Christianity—Orthodox, Coptic, Roman Catholic, or Protestant—might face equal criticism.

As mentioned in Chapter 3, it is wrong to act as if the body, the flesh, is dirty or unimportant. Christianity is not hyperspiritual, and does not overemphasize the spiritual part of our existence. That being said, the Apostle Paul faces accusations, especially from liberal theologians, of importing pagan ideas into Christianity. Hence their continual search for the authentic Jesus or the historical Jesus. Such verses as Romans 7:25, they claim, provide evidence of Paul's Hellenistic reinterpretations of the genuine message of Jesus: "So then, with the mind I myself serve the law of God, but with the flesh the law of sin." Or consider Galatians 5:16–17: "I say then: walk in the Spirit, and you shall not fulfill the lust of the flesh. For the flesh lusts against the Spirit, and the Spirit against the flesh; and these are contrary to one another, so that you do not do the things that you wish." Paul was no Platonist or "innovator with reference to Jesus."[546] He was an apostle of God teaching the pure and true words of God.

What of Christ's exhortation (Mark 14:38) to His disciples just before His death? "Watch and pray, lest you enter into temptation. The spirit indeed is willing, but the flesh is weak." The frailty of their flesh was only too clear, as was the need to sacrifice the flesh for the sake of enduring spiritual blessings. Here is the genuine Jesus: "It is the Spirit who gives life; the flesh profits nothing" (John 6:63). Let us be clear: Jesus was no pagan. Nowhere in all these verses are we justified in presuming the pagan view that spirit is good, but the body evil. The point is that religious notions that originate in the flesh profit nothing. "God is Spirit, and those who worship Him," as Jesus said (John 4:24), "must worship in spirit and in truth." In all these verses the clear view is that this thing called flesh is evil, and opposed to all good, namely, that which is in accord to God and His law. The flesh opposes the spirit. It is our enemy to resist and subdue at every turn.

Christianity teaches the desperate wickedness of the whole man, head and heart. The whole man must change, be born again. At this spiritual rebirth what happens to the physical body? Does it die? Or is the new follower of Christ, the Christian, suddenly able to dodge bullets in a Platonic Matrix? Has he become Nietzsche's Superman? In Romans 8:10, it is written, "And if Christ is in you, the body is dead because of sin, but the Spirit is life because of righteousness." So his flesh is put to death while his mind becomes progressively renewed.

But what exactly is this *flesh* the Bible speaks of? It is clearly more than the material body, the tent of our sojourn (2 Corinthians 5:4; 2 Peter 1:13). For God commands we care for the body and love it (Ephesians 5:29), but the flesh He commands us to hate as an implacable enemy. The physical parts of our body, the atoms, are not evil any more than a tree, or a flea, is evil. Under a divine curse, yes; evil, no. The key is to realize that flesh can refer to the natural fallen nature of man, our body parts made to act as tools, or all our senses tuned to performing sinful acts.

To quote Jonathan Edwards: "Man's nature, being left to itself, forsaken of the Spirit of God, as it was when man fell, and consequently forsaken of divine and holy principles, of itself became exceeding corrupt, utterly depraved and ruined: and so the word flesh, which signifies man, came to be used to signify man as he is in himself, in his natural state, debased, corrupt, and ruined. On the other hand, the word spirit came to be used to signify a divine and holy principle, or new nature; because that is not of man, but of God, by the indwelling and vital influence of his Spirit. And thus to be corrupt, and to be carnal, or fleshly, and to walk as men, are the same thing. And so in other parts of Scripture, to savour the things that be of man, and to savour things which are corrupt, are the same; and, sons of men, and wicked men, also are the same, as observed before. And on the other hand, to savour the things that be of God, and to receive the things of the Spirit of God, are phrases that signify as much as relishing and embracing true holiness or divine virtue."[547]

Christians strive to be God centered in all ways, theocentric as opposed to anthropocentric. Though Christians may err, the general accusation from environmentalists—that Christianity incorporates anthropocentric Greek notions and needs reinterpretation for the new age—is altogether false. As for the actual physical body the Biblical Christian view never denigrates but lifts it up to the heights of glory, even now naming the body of the Christian a sacred temple of the Holy

Spirit (1 Corinthians 3:17; 6:19), a surpassing honor. We are creatures with physical and spiritual natures mysteriously united. The physical nature, like the spiritual, is fallen, and the body is subject to the curse of death, but the body will rise again. Jesus Christ can redeem, and will glorify it. Indeed, by the flesh of Jesus God is reconciled to us.

## Sprouts, Not Meat

Rather than entertaining the now common accusation that pagan heresy infuses Christianity[548] professing Christian Greens should evaluate and reject the heretical stances of the Green Dragon. And then they ought to repent and drive the gargoyles out of the church, cut the idols down, and strive to cast "down arguments and every high thing that exalts itself against the knowledge of God, bringing every thought into captivity to the obedience of Christ" (2 Corinthians 10:5). In fact it is modern environmentalism, not Christianity, which trysts with the worst of classical pagan ideas.

For instance, consider the frenzied insistence on organic food, on vegetarianism, and various such curiosities that pervade the Green movement, increasingly finding a niche in the churches. Organic food sales in America are rising by about twenty percent[549] a year, making this the fastest growing slice of the food market. Now, there is nothing inherently wrong with eating organic foods, even though in doing so one is playing along with harmful cultural myths about nature. But why is this market growing so fast? It is not because organic food is actually healthier, or safer, than most genetically modified crops or food harvested from nonorganic farming—the science supports no such conclusion.

Organic produce contains natural pesticides, often more than in produce grown using conventional pest management.[550] But at least, with conventional produce, one can wash off synthetic pesticide residues. Both conventional and organic foods also contain dangerous toxins; apples contain estragole and apple seeds contain cyanide, coffee contains benzene, arugula and various other greens contain allyl isothiocyanate, grapes contain caffeic acid, red wine contains ethyl carbamate, and so on.[551] Indeed, natural pesticides like these typically account for 99.9 percent of our daily pesticide intake, the synthetic residues for the rest.[552] So when one hears how diet can cause cancer one should realize the same was true for our ancient ancestors. They did not live much shorter lives than we because they ate only organic, but in large part because it was too expensive to eat the quantity and wide variety of clean, fresh produce, organic or otherwise, that we enjoy today. Anything that decreases the intake of *fresh* produce, such as expensive alternatives, inadvertently increases cancer.

The organic market is growing faster than China, partly because the public is conditioned by the idea that somehow Mother Nature always knows best. Whether it is healthier or not, consumers think that organic is better, and feel empowered by the ability to make the "natural" choice. More power to them; if they can afford the expense of organic foods the psychosomatic benefits should not be discounted. But for radical environmentalists there is also a sense of righteousness that comes with eating Green. Try questioning the rationale of organic food at your local raw, whole, or organic market. One journalist recognizes this is "not just akin to doubting the virtues of motherhood, but to reveal indifference to the poisoning of the nation and the fate of the planet."[553]

And now such intolerance is routinely challenging omnivorous diet choices. Theologian Rosemary Ruether writes, "Another shift is toward eating lower on the food chain, greatly reducing meat-eating in favor of vegetable proteins."[554] Because she has some decency she is willing to admit the difficulty of ordering an organic chocolate tofu smoothie in Darfur. So she loosens the Green choke hold just a bit, arguing that only those who have the financial wherewithal "can and should adopt a near-vegetarian diet for the sake of the rights of both animals and other humans."[555] In judging Ruether's preference one should recall which sacrifice was pleasing to God—not the organic vegetarian platter of Cain, but the bloody lamb of Abel.

Do we become wiser than God by deciding that meat eating is ungodly? Princeton University philosopher Peter Singer thinks so. Vegetarian since 1971[556] he states, "[f]lesh taints our meals."[557] Singer is not so softhearted (or softheaded) as Ruether and does not shrink like a wilted violet from the implications of his philosophy. He presses moral boundaries. Since he believes man is just an animal and since under certain circumstances we may justifiably kill animals we may also, under similar circumstances, kill a human being. He believes that "killing a newborn baby is never equivalent to killing a person, that is, a being who wants to go on living."[558] Philosopher Erazim Kohák condenses Singer's views: "every time I refuse to eat the torn bodies of my tortured fellow beings, I save suffering. When I make a hamburger out of Savoy cabbage instead, there is a tiny sliver of suffering less in this world."[559] Vegetarians, writes Kohák, "can look a cow in the eye with a clear conscience."[560] Like Singer, his optimization problem is to minimize pain and suffering, which leads him to vegetarianism.

Kohák is hopelessly anthropocentric when he defines pain by a human experience. Experiments with Kirlian photography suggest trees and other plants somehow sense the loss of limbs, and how one should quantify at least their psychic pain is beyond me. One might insist that since he makes the argument to pain Kohák should justify his slaughter of baby carrots and other vegetables. Is it better to boil them slowly or rapidly, amputate with a knife, or bruise and crush with a hammer?

If not to paganism then, some Greens would have us believe, Jesus leaned toward vegetarianism—though He almost certainly ate beef and fowl. We know He ate fish and, an observant Jew, He ate lamb as required by dietary law. What is more, all evidence suggests He had the duty to cut the lamb's throat before the barbecue, hardly a good vegan practice. We know this because He was the first son, and because, at His crucifixion, He delegated responsibility for Mary to His beloved disciple (John 19:26–27), something He could only do because His father had passed on. Since the Bible allows, even encourages, consumption of animal flesh, the vegetable argument sways merely the most naïve Christian soymilk drinkers.

Those who are not milk drinkers, but have moved on to a meatier spiritual diet, need different poison to disorient them. Guilt and fear are on the menu. Some say Christians should be vegetarian because, as is written in 1 Corinthians 8:13: "if food makes my brother stumble, I will never again eat meat, lest I make my brother stumble." At best it is a contextual misunderstanding, at worst an abusive twisting of Scripture to claim this verse teaches vegetarianism. But at least an appeal is made to Scripture. The discussion here involves food offered to idols. Verse 8 makes clear God allows us full use of all foods: "But food does not commend us to God; for neither if we eat are we the better, nor if we do not eat are we the worse." Later arises the issue of the Christian brother weak in conscience. Paul says he would be willing to renounce meat for their sakes, because of their fear the meat just might have been part of idol worship. It is better to avoid eating if there is the risk of it being an offense to Christians who think it wrong. "I know," writes the apostle of grace, "and am convinced by the Lord Jesus that there is nothing unclean of itself; but to him who considers anything to be unclean, to him it is unclean" (Romans 14:14).

Environmentalists are the modern-day fault-finding Pharisees. They feel they are strong and have the moral high ground. Here lies still another clear signal of the pagan heresy contained in radical environmentalism. From the beginning the self-righteous have made hay with asceticism. These days, the Greener-than-thou do so in terms such as whether one drives a hybrid car, eats certain foods, and treads on the earth in certain ways. In fact, Rev. Jim Ball, the executive director of the Evangelical Environmental Network, an important pillar in the Al Gore-sponsored National Religious Partnership for the Environment, argues for a Green Jesus who would drive something like a Toyota Prius.[561]

What does it take to be a good person, as an environmentalist? It is not enough to plant trees and stop kids from teasing animals at the zoo. It is no longer even enough to go "Green," paying lip service and massive taxes. Tangible sacrifices supposedly prove the extent of true devotion such as using only two squares to clean up after a toilet visit, or driving a hybrid car, or not driving at all. But none of these sacrifices are enough. In an interview with Supreme Master TV, NASA's Dr. James Hansen said, "Be veg, go green, save the planet."[562] The United Nations declares humans should eat less meat to save the planet. Dr. Rajendra Pachauri, the railway engineer cum chair of the U.N. Intergovernmental Panel on Climate Change (IPCC), who with Al Gore shared the 2007 Nobel Peace Prize, says we must cut down meat consumption. "[W]e really," he explains, "have to bring about reductions in every sector of the economy."[563] Pachauri happens to be a close associate[564] of John Holdren, the Science Czar for President Obama. And predictably, Pachauri is a vegetarian and into population control. One wonders if the IPCC will develop a meat patch, like a nicotine patch that food addicts can take, whenever the urge strikes to eat a steak and end the world.

It is an offense to Pachauri, and many other Greens, that humans preserve their lives by eating and drinking dead creatures; it is against their religion. Our need to feed unquestionably shows human frailty, but to the environmentalist it shows in addition how spiritually degraded we are. Do not touch, do not taste, is what they say. Why, they wonder, would any righteous person consume the flesh of dead animals when they could instead consume creatures that live on light, such as plants? How Platonic is that? When we eat plants we are closer to the light; we eat secondhand light! But when we eat meat, we eat secondhand dirt. A similar "higher life" heresy attended the early Church. God has a word for these folks (Colossians 2:20–23): "Therefore, if you died with Christ from the basic principles of the world, why, as though living in the world, do you subject yourselves to regulations—'Do not touch, do not taste, do not handle,' which all concern things which perish with the using—according to the commandments and doctrines of men? These things indeed have an appearance of wisdom in self-imposed religion, false humility, and neglect of the body, but are of no value against the indulgence of the flesh."

All of this Green blather about food is reminiscent of ancient Greek paganism. Pythagoras through Plato were famous vegetarians, arguing that it was immoral to eat animals since they shared with humans a common soul.[565] Professing Christian Greens risk similar odd and spiritually dangerous entanglements. Christianity, Greens tell us, with its crazy ideas about man being the image of God, alienates us from ourselves and the divine. We are always trying to be someone else, so are never able to be at peace, just as we are. Not at all. It is radical environmentalists who are out of touch with the most basic of human needs and emotions. Sister Marilyn Rudy of Earth Harmony in California says, "Food is holy. Let us eat as a spiritual experience."[566] This root of environmentalism is pure paganism, as in this zinger Rudy takes from the sheer joy of gardening: "I realize we are all one piece. I become one with the earth, I am at peace, and know the soil."[567]

One is glad that she attains some satisfaction from gardening. But such a romantic view of nature is no guarantee of a peaceful, or good, soul. Mao Tse-Tung, the biggest monster of the twentieth century, was supposed to be *the* great agrarian reformer. Stalin was the "great gardener,"[568] Adolf Hitler was himself an avowed vegetarian, and his

Deputy Führer, Rudolf Hess, was an organic food and environmentalist fanatic.[569]

The völkisch movement of the 1920s, which foreshadowed the rise of the National Socialists, regarded capitalism, urbanism, rationalism, and science as artificial and evil, alien to the "essence"[570] of the healthy man. In *Liberal Fascism* Jonah Goldberg explains how "Organic food was inextricably linked to what the Nazis then described—as the left does today—as social justice issues."[571] With their various Green rules and regulations, such as a total ban on smoking in public places,[572] there is an unsettling resemblance to the shrill behavior of the twenty-first century Green movement. Now, just as then, the Greens "invoked a nature-romanticism in which closeness to the natural landscape was to give people a heightened sense of aliveness and 'authenticity'."[573]

There have been so many attacks on human consumption of meat that one needs a strong dose of truth to offset the distortions. Meat is supposed to be unhealthy, but God says to eat with gusto (Acts 10:13). Peter Singer says, "[t]here can be no defense of eating flesh in terms of satisfying nutritional needs."[574] But it turns out that good taste, and science, questions this conclusion. Meat is a good source of protein, iron, zinc, selenium, magnesium, calcium and a range of B vitamins. In outstanding recent news a transfatty acid called conjugated linoleic acid (CLA) is listed as an off-the-charts potent anticarcinogen.[575] The world's richest sources of CLA are beef and lamb products. So for your health forget about bioflavonoids, go get a steak … while you still can.

## Der Kommissar

Radical environmentalists want a central role in even the most basic parts of human existence, such as food consumption. They seem to have a fixation on food, so often is it mentioned in environmentalist literature. No one else cares that many environmentalists choose a vegetarian diet. The problem is that they are not content for diet to be a personal preference. A significant trend in the Green movement is the urge to turn Green lifestyle preferences into law, hence a profusion of what have become known as food Nazis.[576] What this means, of course, is coercion. By some accounts, such as those by Paul Ehrlich, the end of the world has already come, but by others, such as the IPCC, we have until 2015 to save the planet. Coercion becomes necessary since we are dealing with an emergency brought on by the moral and ethical failures of humanity.

"Recently I was foolish enough to try to reason with an environmentalist," wrote the Stanford economist Thomas Sowell. "But it became obvious that he had his mind made up and didn't want to hear any evidence to the contrary. The pope is more likely to have read Karl Marx than an environmentalist is to have read even a single book that criticized environmentalism."[577] Karl Marx and money seem also to be obsessions of the environmentalist movement. Though not publicly advertised as a program for one or other form of socialism Greens invariably espouse the same old-fashioned and failed ideas. Professing Christian environmentalists also display untoward affection for economic and political systems that crush the individual and exalt collective action and groupthink. As summarized by Robert Fowler, in *The Greening of Protestant Thought*, "If we want to save the planet, we must turn to an economics that will speed the arrival of sustainability, not growth or expansion."[578] A *New Scientist* article barked that it is "Time to Banish the God of Growth."[579]

Perhaps the main reason for socialism's popularity in the Green movement is that it implies government control over all areas of society. At a recent visit to Oxford University Al Gore said that environmental problems, such as global warming, present as great a challenge to civilization as ever faced. He invoked the spirit of Winston Churchill "who aroused this nation in heroic fashion to save civilization in World War Two."[580] To great applause Gore constantly calls for increased concentration of power in the hands of an intelligentsia, a few good government bureaucrats, who will exercise dominion over the proles. He says the world needs a "global Marshall plan."[581] In other words, we can only be saved by nothing less than a move from the liberty of free and uncoerced decisions, to the irresistible and tyrannical pressure of state power to micromanage the (trillions of) decisions of individuals. "[A]ction must be taken now,"[582] insists the central committee of the World Council of Churches. So the Green program wants not only a change of consciousness, but also of institutions and ways of living.

A second cause for its popularity is that socialism need not always be frowning, as in the former Soviet Union, Red China, or North Korea, but can have varying degrees of smiley faces, Sweden for instance.

With the government in control and management of national real estate and insurance assets, the USA[583] is on the way to becoming the largest smiley face socialist democracy in the world. Smiley face socialism seems charming, because it temporarily offers an illusion of freedom, proficiency, and choice but vests complete control of outcomes in Green rules and regulations. It provides a reality show in which one is free to pull the handle on the slot machine, but all the slots come up the same, guaranteeing results always against your best interests.

"If we wish to preserve a free society," said Friedrich Hayek, "it is essential that we recognize that the desirability of a particular object is not sufficient justification for the use of coercion."[584] But as we have seen, more and more people feel quite comfortable with government coercion in the most mundane decisions, from what they may eat to what they may say. Environmentalists have ranked their priorities and found they prefer their ideas of cosmic justice, though incompatible with preservation of individual freedom. Perhaps they imagine they will be the ones in the driving seat of Green legislation. They think one cannot remake the world without the application of threats and force, and they are fine with that. "We don't want people to use disposable bags. We want people to use reusable bags,"[585] says the Natural Resources Defense Council. And so it goes.

The Bible is an anti-socialist document.[586] Gary North says, "Socialist propagandists for over four centuries have claimed that the Bible teaches socialism, but we have yet to see a single Bible commentary written by a socialist. If the Bible teaches socialism, where is the expository evidence?"[587] But now that a demoralized public has largely swallowed environmental apocalypse, especially via runaway global warming, it is easy to convince some Christians that they have a responsibility to force draconian changes in the lifestyles of their neighbors. This can be made to sound so tender. It is for the sake of the children, after all, whom we all love. But environmentalists use God words while simultaneously denying that God's law in any way provides the definitions, specific content, and intellectual limits of loving my neighbor. Only an abstract idealist, who would never use his hand to pick the pocket of a neighbor, would consider it loving to use the power of the state to rob and force hundreds of millions of his neighbors. But we dare not set aside the eighth and tenth commandments in favor of manmade ideas of social justice, or biospheric justice.

Were their message simply to encourage greater personal generosity with one's own time and property, there would be no quarrel. But environmentalists invariably argue that it is a moral duty to redistribute the property and wealth of others. In *Saving God's Green Earth* pastor Tri Robinson argues that the environment is not a political issue, and Christians can be environmentalists too.[588] But environmentalism is mainly a political project, and it is certainly an ideology religious in nature and opposed to Biblical Christianity. Robinson tries to pull a veil over the Green corpse but unintentionally trips into blasphemy, recently stating to the *Huffington Post*, "Salvation is the heart of the Christian faith—and Christians can no longer ignore a planet that needs saving."[589] Willis Jenkins offers much of the same in *Ecologies of Grace: Environmental Ethics and Christian Theology*. His project is to refocus Christian environmentalism from a formation around the doctrine of creation to one around views of salvation. This change, he says, can help Christian people, and others, find a saving, therapeutic experience in encounters with nature.[590]

Unfortunately Christian advocates for such touching encounters are one with the political Left wing in unvarying advocacy for the socialist political outlook. Basically, the only way to solve pollution problems is to steal from one to give to another. They wish to abolish capitalism, the system in which individuals decide for themselves, free from state compulsion, how they will apply their capital and resources. "Individualism *is* deeply attractive," says Bill McKibben disapprovingly in *Deep Economy*, his shallow book that applies deep ecology to economics. McKibben's latest activist tool is a group, 350.org, that calls for governments of the world to take us back to a time when CO2 allegedly comprised 0.035 percent of the atmosphere. McKibben says, "It's the most important number in the world … the line between habitability on this planet and a really, really desolate future."[591] In *Earth in the Balance* Al Gore struggled throughout to balance his Jeffersonian support to individual rights with his visceral lust for more direct approaches to governance. He admits problems with socialist excesses, writing for instance, "the very worst environmental tragedies were created by communist governments, in which the power of the state completely overwhelms the capabilities of the individual steward."[592]

Fifteen years later Gore has surrendered, perhaps through frustrated impatience, to more totalitarian impulses. In his speech to the Nobel Committee he said, "We must abandon the conceit that individual, isolated, private actions are the answer. They can and do help. But they will not take us far enough without collective action. At the same time, we must ensure that in mobilizing globally, we do not invite the establishment of ideological conformity and a new lock-step 'ism.'"[593] His passing Orwellian wink to the earlier unpleasantness of twentieth-century gulags does nothing to steady the racing pulses of wannabe totalitarians the world over. It is not liberal policy to apply state force as a primary tool for effecting moral change in society.

Gore's vision of humans bound to the single-minded imperatives of the collective is as unattractive as it is unworkable. In this vision, something akin to that voiced by Maurice Strong, heads of state will act on imperatives of the new Green central organizing principles, as members of an international comintern, rather than as servants of their people. And the people must pretend to be content as they blankly pull on that one-armed bandit whose answer is always the same.

## Green with Envy

"In the past 150 years (at least since Marx)," wrote the President of the Czech Republic, Václav Klaus, "the socialists have been very effectively destroying human freedom under human and compassionate slogans, such as caring for man, ensuring social equality, and fostering social welfare."[594] All the solutions to environmental collapse offered by politicians require tough government controls on individuals. The new slogan is planetary salvation, and Green leaders think Christians will wiggle and fawn like pleasure-drunk puppies because of the God-talk. Professing Christian environmentalists are the useful idiots of the Green Dragon. They affirm "love to neighbor" or "creation stewardship," which makes it sound as if they affirm the Bible, keeping dubious Christians constantly confused. Then, while the baffling slogans flash like machine guns in the minds of their hearers, Greens hustle in the fads of the political left as they progressively rob and enslave their brothers.

"The richer we get," says Bill McKibben, "and the more meat we want to eat as a result, the worse the pinch gets."[595] Such sophism masks the motive of envy. The goal is to make the one being envied feel guilty. If only I did not own stuff there would be no envy, and so I am responsible for the sin and sufferings of the one who envies. If only I did not eat meat; if only I did not drive a car; if only. This is pure Marxist guilt manipulation. But one need not be a conscious commissar to fall into unbiblical thinking. In the book *Serve God: Save the Planet*, Christian physician Matthew Sleeth writes, "It is not our spiritual longings but our material desires that keep us from a right relationship with God."[596] Covetousness and lust are indeed terrible sins. But throughout his book Sleeth argues that it is not only the excessive desire, but the actual possessions that are a spiritual problem. Jesus says exactly the opposite, "Not what goes into the mouth defiles a man; but what comes out of the mouth, this defiles a man" (Matthew 15:11).

Facts should be important. And the facts show that the world, as a whole, is healthier, wealthier, safer, cleaner, and eating better than ever before.[597] But, say the Greens when challenged by the facts, this cannot continue. They adopt the Malthusian mythology of a zero-sum game. If you are eating well, it is only because someone else is eating poorly. And sometime soon, because of these imbalances, the world will end. Were they serious about ending hunger they would try to understand why hunger left those nations once dominated by evangelical Christianity. It is not that Americans, for instance, are taking food out of the mouths of the world's poor. The United States is a net exporter of food and has fed—for free—about 3 billion people in 150 countries[598] in the past 50 years.

In *Affluenza: The All-Consuming Epidemic,* filmmaker John de Graaf and Duke Economics Professor Thomas Naylor write, "It's as if we Americans, despite our intentions, suffer from some kind of Willpower Deficiency Syndrome, a breakdown in affluenza immunity."[599] Even though the end is near Americans willfully shut their eyes so they can keep stuffing their faces. In *Live Simply: Let Others Live,* Edward Echlin argues that we must downsize, and in so doing we will not only save the planet but achieve some measure of peace. "Living simply," says he, "is ecumenical. Sustainable sufficiency unites people of different 'hopes and beliefs', and demonstrates the great potential of faith to make a difference to our world."[600]

Christianity teaches us that the truly abundant life is one of restraint on sinful desires because of an overriding love for Christ. But gospel power, not legislation to change global consciousness, is what restrains sin. Environmentalism teaches that the abundant life is achievable via self-discipline and restraint on material desires. We are basically animals, they say, and animal behavior may be modified by laws and re-education. Not so; it is false *works* religion that defines righteousness by discipline and bodily denial, by compulsion, if necessary. It misses the Biblical emphasis. Hearts do not usually change by legislation or education, yet this is exactly what the words of environmentalists imply.

Professing Christian Greens are correct when they say true Christians have an outlook opposed to mindless consumerism. One can applaud this observation. Mad pursuit of wealth is termed vanity in the Scriptures, as are flashy jewels and clothing, and lusts of the flesh. Dr. Martyn Lloyd-Jones preached, "[T]he life which is lived by the Christian must be an essentially different life from that of the man who is not a Christian."[601] But there is equally no intrinsic spiritual virtue in downsizing, an idea which enthralls professing Christian Greens. Matthew Sleeth uses Mark 8:35–36 to argue a "downwardly mobile"[602] life pleases God. Those verses say, "For whoever desires to save his life will lose it, but whoever loses his life for My sake and the gospel's will save it. For what will it profit a man if he gains the whole world, and loses his own soul?"

Sleeth believes this same verse, and verses from Proverbs, mean that people who buy SUVs for their safety record are sinning by not putting their security in the hands of God. The planet is dying he says. It is on life support. We must thus leap out of the SUV, highly rated for safety, into a flimsy 1,300 lb Tata Nano. To stay in the SUV, he argues, is to distrust God. But God has nothing against our using sanctified sense, and actually requires it. Throughout Scripture God makes plain that the law of self-denial and suffering for Christ's sake does not negate the duty to preserve our lives by lawful means. So it is perfectly reasonable to prefer a safer vehicle. One wonders whether Sleeth considers it a lack of faith to use safety belts, or to wash fruits and vegetables. Sleeth's lazy and manipulative Scripture twisting is not an isolated aberration among environmentalists.

In Bill McKibben's ideal world, "No one gets superrich … but plenty more people get to do something lovely."[603] Here is the rub: McKibben wants enlightened experts, perhaps people like himself, to approve the lovely for someone, and force everyone else to support it. Rather than follow God's law Green commissars invent their own. This is the definition of Pharisaism and legalism. Consider the law to keep the Sabbath holy (Exodus 20). If our society did obey presumably our environmental impact would immediately drop, if not by fourteen percent (one day in seven), by some level that is not insignificant. But instead of Biblical law Greens adopt their own. By adopting the socialist analysis of property and envy, Greens risk conflict with Biblical teaching, for instance that diligence breeds wealth (Proverbs 13:4) while the lazy should and will go hungry (19:15). Despite all the best efforts, and utopian legislation, there will always be poor in this world as Jesus teaches in Matthew 26. And mind that He said this in response to disciples offended by the conspicuous consumption of a woman who poured precious perfume on His head. Judas was so offended at the waste that he had murder in his heart. Jesus rebuked the indignant ones telling them to worry about their hearts rather than with what a woman did with her own wealth.

Perhaps McKibben would be a kind slave master, but there are many others who, if given the power and opportunity, would enjoy dictating what people can eat, watch, say, and believe. They will always justify it as loving or good; they already do, as in: you may not use regular light bulbs, because we have to save the planet. Or, you may not eat meat, because we have to save the planet. Or, you may not have this child, because we have to save the planet, for the children. While on the campaign trail in 2008 Barack Obama said, "We can't drive our SUVs and, you know, eat as much as we want and keep our homes on, you know, 72 degrees all the time … and then just expect every other country is going to say OK."[604]

Writes Peter Singer, "An environmental ethic would find virtue in saving and recycling resources, and vice in extravagance and unnecessary consumption." This is all good and well, were it on a voluntary, individual level, but environmentalism preaches abstinence from voluntary measures, increasingly preferring the totalitarian solution of government compulsion. The latter is unavoidable since someone needs to decide what is necessary and what is unnecessary. If individuals are robbed of the ability to decide, some special government expert will decide for them. Were he in charge Singer makes plain that governments would ban "motor car racing and waterskiing" but "[c]ycling and windsurfing" would be acceptable. In the realm of diet he might allow "caviar or truffles," but forbid food for common folk, namely "beef, pork, and poultry."[605]

## Watermelons

Christopher Monckton, 3rd Viscount Monckton of Brenchley, recently said, "Greens are too yellow to admit that they are really red."[606] It is difficult to disagree with this amusing statement. Consider Evo Morales, the Bolivian Presidente, who quotes Al Gore with what one can only assume is admiration: "The world is suffering from a fever due to climate change, and the disease is the capitalist development model."[607] Morales shows that it is possible, even common, for reds to turn Green. Then here is more Gore, who says from one side of his mouth that he is for free markets, but from the other side defines freedom by obeisance to Green central organizing principles. His idea of freedom is uncomfortably close to the former Soviet idea of peace; namely, being swallowed up by the collective. "I bring you good news from the U.S.," Gore said on July 7, 2009 in Oxford at the *Smith School World Forum on Enterprise and the Environment*. There he announced that the 2009 U.S. Congressional climate bill (the one James Hansen called a Ponzi scheme) will help bring about "global governance."[608] Gore merely echoes the sentiments of former French President Jacques Chirac who, almost ten years ago, said the Kyoto Protocol represented "the first component of an authentic global governance."[609] Canadian Prime Minister Stephen Harper, now a believer, once labeled Kyoto "essentially a socialist scheme to suck money out of wealth-producing nations."[610] To think we believed environmentalism was about protecting natural resources!

In 2009 Dany Cohn-Bendit, co-president of the group European Greens–European Free Alliance, transformed at a stroke from Dany the Red into Dany the Green–a Watermelon, Green on the outside, Red on the inside. He is surfing the fashionable political wave of environmentalism onto a deeper Red tide. "We have a project for Europe, an idea—the ecological transformation of our way of production and our way of life." It is sickly sweet that peace, love, and justice are what world-destroying socialists are always saying they are about. Says Dany the Green: "It's for the survival of mankind."[611] Self-described socialist activist Tom Athanasiou, director of U.S.-based EcoEquity, wrote "[E]nvironmentalism is only now reaching its political maturity."[612] What he means is that there is a wonderful convergence of Green concerns with Red concerns, where true dialogue can make a lasting impact. In the U.S. the President's short-lived "green jobs czar," Van Jones, has said, "I'm willing to forgo the cheap satisfaction of the radical pose for the deep satisfaction of radical ends." He self-identifies as a "communist"[613] and has discovered in environmentalism a means to satisfy his need for both the radical pose and radical ends. Environmentalism serves all the policies he already believes in.

While the ecosocialist current within the Green movement is advertised as one of many, appearances suggest that it is a fundamental pillar of the Green movement. It is hard to avoid this conclusion given the profusion of Red communist party hammer and sickle flags at the United Nations Climate Change Conference in Copenhagen. For what does it matter that Al Gore claims not to be a socialist when he adopts in practice positions that are inherently so? Saving the planet may be well-intentioned, and no one imagines people like Gore want to harm anyone, but a cure that needs control of every important part of life, fairly described as totalitarian, is worse than the disease. A 1930s-era National Socialist slogan proclaimed, "Nutrition is not a private matter!"[614] That is tame compared to what Green zealots implement today and offer for the Green future. Sadly, only a change of names and places separates most of what the Green movers and shakers are saying from the professing Christian wing of the Green movement. Except for a few God words Christian Greens sound much the same as their secular counterparts.

## Heart Matters

The world's longest war smolders on at the border between the Koreas. But Kim Jong-Il wants peace. John Lennon earnestly dreamed of all the people living in peace. "We all came here for peace," said Christine to her friend Jim, at the communist experiment called Jonestown. "I tried to give it to you,"[615] replied Rev. Jim before serving cyanide-laced Kool-Aid. We all want peace.

Environmentalism promises peace. Writes Robert Fowler, "People will come to appreciate that to go slower, longer, and restore a sense of peace in their lives can be a wonderful thing." [616] He is talking about the benefits of being downwardly mobile, of embracing Green asceticism. Contemplation of nature is desperately important to the environmentalist for from it he draws a sense of the transcendent. After a year-long voluntary experiment in eating only food grown a short distance from where he slept Bill McKibben reported that he had found a measure of peace: "That good taste was satisfaction. … In my role as eater, I was part of something larger than myself that made sense to me—a community. I felt grounded, connected."[617] The theme of the Roman Catholic Church's 2010 World Day of Peace, marked on January 1, was "If You Want to Cultivate Peace, Protect Creation."[618]

To learn the true way of peace one is better advised to read Puritan classics like *The Rare Jewel of Christian Contentment* by Jeremiah Burroughs, or *The Art of Divine Contentment* by Thomas Watson. Both treat the subject of contentment based on Paul's words in Philippians 4:11, "I have learned, in whatsoever state I am, therewith to be content." That is to say, whether I have many or few possessions, whether rich in money or poor, I can be content. Watson says, "Contentment lies within a man, in the heart; and the way to be comfortable, is not by having our barns filled, but our minds quiet."[619] Some people try to find peace in the sin gluttony. They seek comfort of life in having much, or in owning the latest DVD or gadget, or in having more than others. Professing Christian Greens are expert at pointing out this splinter, which they call world-ending conspicuous consumption, but fail to see the beam that pins them to the ground. We should do as they say, not as they do (Matthew 23:3). They are so married to the green Earth that they are unable to rise heavenward, which is the only way for a Christian to go. Says Thomas Watson, "When a stone ascends, it is not a natural motion, but a violent; the motion of the soul heavenward is a violent motion."[620]

Those who chase Green peace harbor disorder in their minds because their very existence is pollution to the Earth. Their flawed theory is an alternative religion that elevates Earth and animals and creates an unquenchable dissonance in their heart. People fear because they cannot make sense of their world and are alienated from God. They feel trapped in a fatalistic death spiral from which escape is impossible. The Green Dragon offers meaning through action. It is an empty lie. God's Word points out the only way to peace: Jesus said to His followers (John 16:33), "These things I have spoken to you, that in Me you may have peace." Any other route is hopeless, for "'There is no peace,' says the LORD, 'for the wicked.'" (Isaiah 48:22). Indeed, "the wicked are like the troubled sea, when it cannot rest, whose waters cast up mire and dirt" (Isaiah 57:20).

They torture themselves (and others) with their troubles. Rather than being content, they are full of envy and bitterness of soul, forever looking at what others are doing, forever resenting that others do not care as they do, forever envying. They gather a personal sense of worth and sometimes superiority from their Green vision of the world. And they are not going to give that vision up easily. Greens will discover that "envy is rottenness to the bones" (Proverbs 14:30). The psychology of the Green, especially the professing Christian Green, guarantees a life of unhappiness, for there will always be environmental challenges because of sin and the curse (Genesis 3:19). This environmentalist resentment, anger even, is like an addiction. Like an alcoholic, the more he drinks, the more he thirsts. There can only be a constant hunger for an elusive taste of true spirituality, and no rest, no lasting contentment, no peace.

## Sin

Environmentalists define sin not by the law of God, but by Green central organizing principles. For instance, in *Collapse,* Jared Diamond produces a satellite composite of the world at night showing how brightly lit are the USA, Western Europe, and Japan, and how sparsely lit is the rest of the world. His message is that bright lights are bad for the environment, examples of glaring overconsumption. But one can get a thoroughly different impression from the image. Except for South Africa and the glowing upper African eyebrow the rest is the dark continent. A stark line on the face of the earth separates India from communist China, India basking in light. Most dramatic is the line between the two Koreas. In the north, total darkness. The south is like a riot of color splashed on a celestial canvas, a light on a hill. South Korea, like Japan, seems a perpetual construction site. Yet daylight satellite images show, relative to the North, South Korea is green and verdant. North Korea, on the other hand, looks like a blighted wheat stalk, much of it brown and burned over. The communists blame it on the Americans, on the drought, whatever. The destructive effects of drought seem curiously to follow the line of the demilitarized zone between North and South. On one side green and growing, on the other side blasted. Night satellite images show almost no lights in North Korea. It looks peaceful; there is no movement. Though perhaps horrified by North Korea's oppression of its people, environmentalists admire the darkness, ascribing a quiet dignity to the people who consume less. But the stillness is that of the empty womb and the whitewashed tomb.

One might wonder, were they to play Diamond's game, why Kim Jong-Il has not yet won the Nobel Peace Prize. The little man has been ripped off far too long. In President Kim Dae-Jung South Korea received its first Nobel Peace Prize in 2000 for doing nothing. It was Kim Jong-Il, the North Korean tyrant, who showed restraint and preserved the peace, the other Kim being a toothless puppet. Perhaps the Nobel selection Committee confused the two Kim's because their names look so alike? An avid golfer, Kim Jong-Il routinely shoots three or four holes-in-one, according to North Korean state media.[621] Reasonable odds of doing this in one round are around 1 in 2,560,000,000,000,000,000 or the equivalent of being struck by lightning over 100 thousand times a second during a typical three hour game.[622] This explains Kim's hair problems.

Kim is a sensitive man of culture; he has composed six operas and enjoys staging elaborate musicals.[623] Kim is certainly more worthy than Al Gore, for he has put talk into action, turning North Korea into a Green utopia. The communists follow all the policies that western Greens pine for, and then some. They control the minute details of life. They have socialized not only medicine but intimacies of life, and science experts decide the most sustainable path for the nation. From satellite images one can see their carbon footprint is extremely low. Almost all the people live a life of downward mobility. A North Korean gulag escapee noted, "Once in a great while we would have a fish. We could only catch these when the guards weren't around, because fishing was banned at Yodok. The guards—straight faced as ever—said that the rule existed to protect the environment."[624] In addition, Kim has consistently developed weapons to protect the environment instead of investing in polluting industries. Their nuclear missiles have served as a deterrent on American plans for mass exploitation of Alaska. According to the North Korean government they have "consistently directed great efforts to the land management and environmental protection in order to prevent global warming."[625] Why no Nobel?

But seriously, North Korea reminds one of what regimes look like when they expel freedom of the individual. They did it to "save" just North Koreans, but modern Greens want to "save" the entire planet. If a man is a slave to sin he is not able to escape other forms of slavery such as blind devotion to a red, or Green, ideology. Today's educated Westerners might find North Korean brutality revolting, but they are susceptible to subtler forms of ideology, brought in through the environmentalist Green door. Professions of nature love are no comfort. As Staudenmaier and Biehl discovered, "The record of fascist ecology shows that under the right conditions such an orientation can quickly lead to barbarism."[626] At any rate, Green or red, they underestimate or ignore the power of sin in man's heart.

Christians can find common ground with environmentalists by stressing the flawed nature, or brokenness, of man. But the environmentalist is at heart a utopian, believing in unrealistic solutions to (sometimes nonexistent) problems. The crisis of man's depravity seriously cramps the style of environmentalists, as it has ever done to evolutionary humanists. Sin and macro-evolution are incompatible, for the reality of sin implies that education and re-education are powerless to change what is a heart problem. Our ancestry is not animal, but divine. Christianity is not romantic, it is realistic. It sees man, and the world, as they are, without sugarcoating. The environmentalist thinks that it is within the power of man to change his fundamental nature, but Christianity says this is impossible. Yet utopians will still try, as in the former Soviet Union, communist China, North Korea, even revolutionary France. First human character was to change through education, and then, when that failed, through political pressure, and finally violence. How easily good intents go awry. Al Gore does not learn from the past, writing, "A change in our essential character is not possible without a realistic hope that we can make change happen."[627]

Such efforts will always fail because man is not merely flawed but completely broken; dead in trespasses and sin is how the Bible characterizes it. Why? Because man's basic problem is ethical not environmental. Man is a sinner. Even atheists such as Václav Klaus recognize our current danger as reminiscent of past horrors. He writes, "Mankind has already accumulated tragic experience with one very proud intellectual stream that claimed that it knew how to manage society better that spontaneous market forces. It was communism and it failed, leaving behind millions of victims. Now, a new -ism has emerged that claims to be able to manage even nature and, through it, people. This excessive human pride–just like the previous attempts –cannot but fail. The world is a complex and complicated system that cannot be organized according to an environmentalist human design, without repeating the tragic experience of wasting resources, suppressing people's freedom, and destroying the prosperity of the whole human society."[628]

## Conclusion

In ancient Greek myth the dragon Python, son of Gaia, guarded the oracle at Delphi, the name fittingly related to δελφός (delphos), meaning womb. It was here that worship of Mother Earth was practiced, a way spurned by the ones called Christians. Today a new monster assaults humanity, a powerful secular religion in the form of a great Green Dragon, a man-eater. Jesus said, "Beware of false prophets, who come to you in sheep's clothing, but inwardly they are ravenous wolves. You will know them by their fruits. Do men gather grapes from thornbushes or figs from thistles? Even so, every good tree bears good fruit, but a bad tree bears bad fruit. A good tree cannot bear bad fruit, nor can a bad tree bear good fruit. Every tree that does not bear good fruit is cut down and thrown into the fire. Therefore by their fruits you will know them" (Matthew 7:15–20).

The fruits of the Green Dragon are not good, but evil. The words spoken in support of the Dragon are not soothing, but terrible. Humans are urged to surrender as many liberties as judged fit to save the world, which is pretty much all liberty that makes life worth living. But the world needs no saving. Satan has long resented his demotion and fancied himself divine. He has been cast from God's pure presence. The promotion of mankind, in the form of Jesus and His followers, must be too much to bear. He has therefore, through the ages, assaulted mankind. In ancient times it was through Gaia, the Earth goddess, that he led people to Green hell. In our times the Green Dragon does as well. Western Civilization, which once was Christian, is today chiefly pagan, and environmentalism is its most prominent religious expression.

Environmentalists say we need a new central organizing principle of life, namely salvation of the planet. This is an injury to truth—the planet does not need saving. False ministers with no or low faith in gospel vigor have penetrated the Church with a mist of lies. They think that action comes from the bustle of man, and so turn to politics as a solution to their problems. Economics and politics are not the problem, or the solution. The problem is greater even than the change of consciousness, and reinvention of religion. The difficulty is a heart problem. Western Civilization is daily committing suicide, by aborting itself, ignoring the law of God, and binding itself in Green chains. What of professing Christians? It is not clear many believe, as they still did at the American Revolution, that "[t]he truths of the gospel are the most pure, its motives are the most noble and animating, and its comforts the most supporting to the mind."[629] Our culture is arguably as enslaved as were the ancient people chained by Gaia and Python. Only Christ can bring liberation. The Church needs to recover the "Biblical persuasion that the Christian is a pilgrim on the road to a heavenly city, that the social structures of the earthly city do not represent the coming of God's kingdom, and that the vision of a society should be to direct man to what is transcendent."[630]

But transcendence is found nowhere else except in the person of Jesus Christ. Environmentalism is a powerful rival religion, ancient as Eden, old as the ashes, against which Christians are permanently at war. Professing Christian Greens, and their secular counterparts, try to inculcate a sense of overwhelming crisis brought on by the human condition. They offer salvation if only we prepare collectively to impose radical limits in social, political, and economic liberty, and therefore even religious liberty. The problem is life; they think they can cure original sin. These *faux* Christians must be invited to turn from their inner pagan. They do not mention Jesus, except to use as a tool to manipulate ignorant or naive Christians into obeisance to unbiblical principles. But Jesus is *the* central organizing principle of everything. Without Him there is no happiness, no peace, and no salvation. Christians will reject totalitarian lures, for Christ imbues His followers with a spirit of liberty, not lordship over others. And this is why Christians are not helpless in the great spiritual battles of the day.

According to reliable media sources at the *Washington Post* followers of that Way of Jesus Christ are "largely poor, uneducated, and easy to command."[631] But this is only true in part. From the beginning most of the Christians were largely poor and uneducated. Even today, the global center of Christianity is no longer in the West, but is in Africa and the East, where it is progressively transforming those impoverished cultures as it did centuries before among the barbarians of Europe. Today it is the European races that are descending again into barbarism.

It is true that Jesus enjoyed no formal education and mingled with the lowest classes, including tax collectors and prostitutes. He preached a gospel of repentance to them, a gospel at once both cerebral and spiritual. In an astonishing proof of the power and grace of God He—a carpenter—stood and taught in the Jewish Temple, the seat of rabbinical learning for centuries. And when He taught it was with such enormous erudition that He silenced the best arguments of the scholars. The Bible contains accounts of some of his interactions: "And the Jews marveled, saying, 'How does this Man know letters, having never studied?'" (John 7:15).

It was well that the Jews marveled, for Jesus was poor and, by the standards of the day, considered uneducated. He did not deny it or minimize the absence of human scholarship. He openly admitted it, going so far to explain that the source of His great learning and brilliance was not of man, but purely divine. He said, "My doctrine is not Mine, but His who sent Me. If anyone wills to do His will, he shall know concerning the doctrine, whether it is from God or whether I speak on My own authority" (John 7:16–17).

Unlike the Apostle Paul, who sat at the feet of the great Gamaliel, Jesus did not learn from the scholars. He explained this to the incredulous multitude: "The Spirit of the Lord is upon Me, because He has anointed Me to preach the gospel to the poor; He has sent Me to heal the brokenhearted, to proclaim liberty to the captives and recovery of sight to the blind, to set at liberty those who are oppressed; to proclaim the acceptable year of the Lord" (Luke 4:18–19). His authority and power stemmed from God alone. So perhaps those words of the reporter from the *Washington Times* were an unintentional understatement; a hit piece that was ironically on mark. Godless men lift hands against God and God's people. But the strong right hand of God upholds His people and will cut off the wicked.

Christians are to believe in simplicity, but not be simple-minded (Romans 16:19). They are fools for Christ, but not fools (1 Corinthians 4:10; Psalm 14:1). He has sent Christians into a world where the weak and naive are easy prey (Matthew 10:16): "Behold," said Jesus, "I send you out as sheep in the midst of wolves. Therefore be wise as serpents and harmless as doves." What can these weak people do? First and foremost, they should pray, preach, and live the gospel. Certainly not become soft targets, easily captured by vain philosophies. Nor do they avoid love because society revels in perverted lust; they do not abstain from food or possessions because society loves gluttony; they do not avoid concern for a healthy environment because of environmentalism.

In 1856 two young girls from the Xhosa tribe in South Africa reported that on the banks of the Gxarha river mysterious ghosts appeared and taught them to spread this message: "We are people who have come to order you to kill your cattle, to consume your corn, and not to cultivate anymore."[632] Rewards would follow obedience, promised the floating spirits, first a resurrection of ancestors, victory over enemies, and finally rich peace and prosperity. The elders at the kraal laughed at the story. But the witch doctor Mhlakaza believed the ancestors had spoken and the Xhosa had to show faith by sacrificing their most treasured possessions. Mhlakaza used his lore of science and superstition to convince paramount chief Kreli to order all the amaXhosa to obey the ancestors. Many were unwilling but societal pressure, mass hysteria, and threats of violence won the day. After many false starts the deniers, the Amagogotya,[633] were coerced into silence; the lowing and bleating died down. Just over a year later the Xhosa had learned the lesson of their folly, but by then they were short 400,000 head of cattle and 40,000 Xhosa had starved to death.[634] By 1859 the Xhosa population was seventy-five percent smaller than three years before.

The strange compulsion to societal suicide is a dogma in the Green movement, moving ever closer to the edge of a cliff. And they want people to step over the cliff to show solidarity with the new organizing principles. This madness is reminiscent of the national suicide of the amaXhosa. Like Mhlakaza, false prophets promise salvation if only we will destroy the means of maintaining our civilization. No more carbon, they say, or the world will end and blessings cease. Rather than scorning the pagan prophets a similar mass hysteria infects many professing Christian churches. Though the boat still moves forward, the ebb tide of Biblical Christianity is plain. Western Civilization has never known greater prosperity than at the present, but rejection of the Christian foundations of our prosperity is near total in many denominations. Pagans of all stripes now offer their rival views of salvation, all of which lead to death. As it says in Proverbs 8:36, "But he who sins against me wrongs his own soul: all those who hate me love death." And have no doubt, joining voices with the Green Dragon is to sin against Christ.

Christians must not join with these forces, but flow against the ebb tide. Yet too many Christians behave as if they believe the world is Satan's, so they run and hide. The struggle they find too warm. The chariot wheels drag heavily. It is daunting to swim against the current. But we recall that only dead and dying fish swim downstream—living ones go against the flow. The Apostle James (James 4:7) tells believers the devil is the one that should be running; Christians are not at the mercy of the devil and his servants. They are to resist the devil, not retreat. And they do so first by preaching the gospel, the Word of life to a perishing world.

In the name of Christ we uphold the authority of the Word of God. He offers life and dominion. The Green Dragon, as in Eden, offers disintegration of society and death. Christians must resist Green overtures to recast true religion, nor allow themselves to be prey for teachers of pagan heresies (2 Peter 2:1–3). Contemporary theologians have tried to tame the truths of Christianity, but end up consumed by crushing spasms of pagan environmentalism. There is no need to do so, for the head of the serpent will yet be crushed. As J. Gresham Machen wrote one hundred years ago, "We have today the entrance of paganism into the Church in the name of Christianity. But in the second century a similar battle was fought and won."[635]

# Endnotes

[1]Denis Hayes and Daryl Hannah, "Fame Exchange," *Delta Sky Magazine*, April 2009, p. 69.

[2]Earth Day Environmental Gallup Poll, online at http://www.gallup.com poll/117769/Outlook-Environmental-Quality-Improving.aspx, viewed 6/16/2009.

[3]Pew Global Attitudes Project, online at http://pewglobal.org/reports/pdf/256.pdf, viewed 5/12/2008.

[4]Bjørn Lomborg, *The Skeptical Environmentalist: Measuring the Real State of the World*, 1st ed. (Cambridge: Cambridge University Press, 2001), p. 3.

[5]Lomborg, *Skeptical Environmentalist*, p. 4.

[6]Al Gore, *Earth in the Balance: Ecology and the Human Spirit*, 1st ed. (Boston: Houghton Mifflin, 1992), p. 223.

[7]Jonah Goldberg, "Starry Enviro? The Cases of Bjørn Lomborg and Galileo," *National Review Online*, January 13, 2003, online at http://article.nationalreview.com/267626/starry-enviro/jonah-goldberg, viewed 6/23/2010.

[8]Anyone interested in comparing the claims of the Litany against the real state of the environment is referred to books by Lomborg, especially *Skeptical Environmentalist*, and Julian L. Simon, especially *The Ultimate Resource 2* (Princeton: Princeton University Press, 1996) and *The State of Humanity* (Cambridge, MA: Blackwell, 1995).

[9]Robert L. Dabney, "Positivism in England," in *Works*, vol. 3, edited by C. R. Vaughan, (Richmond, VA: Whittet and Shepperson 1892), p. 32.

[10]Paul Ehrlich, cited in Alastair Sweeny, *Black Bonanza: Canada's Oil Sands and the Race to Secure North America's Energy Future* (Ontario: John Wiley & Sons, 2010), p. 234.

[11]"The Cooling World," *Newsweek*, April 28, 1975, p. 64.

[12]Harold M. Schnuck Jr., "Climate Changes Endanger World's Food Output: Scientists View Global Climate Changes as Threat to World's Food Output," *New York Times*, August 8, 1974, p. 35.

[13]"Worrisome CIA report: Even U.S. Farms May Be Hit By Cooling Trend," *U.S. News & World Report*, May 31, 1976, p. 56.

[14]"Next Great Deluge Forecast By Science: Melting Polar Ice Caps to Raise the Level of Seas and Flood the Continents," *New York Times*, May 15, 1932, science section, p. 4.

[15]"Rising World Heat Seen Leveling Off: Northern Hemisphere, After Long Warm Spell, Is Cooling Off, Swedish Expert Says," *New York Times*, August 7, 1952, p. 23.

[16]"Menace of a New Ice Age to be Tested by Scientists: Indications in Arctic That Have Caused Some Apprehension—MacMillan Expedition Will Leave for Greenland Next Month to Study Recent Movements of Glaciers," *New York Times*, June 10, 1923, amusements section, p. 3.

[17]"Prospects of Another Glacial Period: Geologists Think the World May Be Frozen Up Again," *New York Times*, February 24, 1895, p. 6.

[18]"This Climate of Ours: Why These Open Winters and Temperate Summers? The Geological Evidence of the Alternate Prevalence of a Semitropical Atmosphere," *New York Times*, June 23, 1890, p. 5.

[19]Gore, *Earth in the Balance*, p. 245.

[20]Eusebius, *Chronicles*, I.1.1:261.
[21]Julia Marvin, "Cannibalism as an Aspect of Famine in Two English Chronicles," in *Food and Eating in Medieval Europe*, edited by Martha Carlin and Joel Thomas Rosenthal (London: Hambledon & London, 2003), p. 73.
[22]Stéphane Courtois, et al., eds., *The Black Book of Communism: Crimes, Terror, Repression* (Cambridge, MA: Harvard University Press, 1999).
[23]Thomas Hobbes, *Leviathan*, edited by A. R. Waller (Cambridge: Cambridge University Press, 1904), Part I, Chapter 13, p. 84.
[24]Otto Bettmann, *The Good Old Days—They Were Terrible!* (New York: Random House, 1974).
[25]Nicholas Wade, *Before the Dawn: Recovering the Lost History of Our Ancestors* (New York: Penguin, 2006).
[26]Lomborg, *Skeptical Environmentalist*, p. 50.
[27]Quoted in Linda D. Kozaryn, "DoD Tackles El Nino of Cyberspace," Department of Defense, July 2, 1998, online at http://www.defenselink.mil/news/newsarticle.aspx?id=43215, viewed 7/2/2009.
[28]Jean-Jacques Rousseau, *Emile* (Whitefish, MT: Kessinger Publishing, 2004), p. 3.
[29]Francis A. Schaeffer, *How Should We Then Live? The Rise and Decline of Western Thought and Culture* (Wheaton, IL: Crossway Books, 2005).
[30]Georgius Agricola, *De Re Metallica* (Mineola, NY: Dover Publications, 1950).
[31]Mark D. Tooley, "Breaking the P.C. Law," *FrontPageMag.com*, April 5, 2007, online at http://97.74.65.51/readArticle.aspx?ARTID=26027, viewed 6/23/2010.
[32]Friedrich Wilhelm Nietzsche, *Thus Spake Zarathustra*, translated by Thomas Common (New York: Boni and Liveright, 1917), p. 157.
[33]Thomas R. Malthus, *An Essay on the Principle of Population* (Oxford: Oxford University Press, 1999), p. 61.
[34]Nicholas Werth, "The Great Famine," in *Black Book of Communism*, ed. Courtois et al., p. 165.
[35]Jean-Louis Margolin, "China: A Long March into Night," in *Black Book of Communism*, ed. Courtois et al., p. 495.
[36]Diomidis Spinellis and Panagiotis Louridas, "The Collaborative Organization of Knowledge," *Communications of the ACM*, 51(8):68–73, August 2008, online at http://doi.acm.org/10.1145/1378704.1378720, viewed on 7/2/2009.
[37]Dennis M. Wilkinson, "Strong Regularities in Online Peer Production," *Proceedings of the 9th ACM conference on Electronic commerce*, Chicago, IL, 2008, pp. 302-309.
[38]Simon, *Ultimate Resource 2*, p. 380.
[39]"In October 1838, that is, fifteen months after I had begun my systematic inquiry, I happened to read for amusement Malthus on Population, and being well prepared to appreciate the struggle for existence which everywhere goes on from long-continued observation of the habits of animals and plants, it at once struck me that under these circumstances favourable variations would tend to be preserved, and unfavourable ones to be destroyed. The results of this would be the formation of a new species. Here, then I had at last got a theory by which to work." Charles Darwin, *The Autobiography of Charles Darwin* (London: Collins, 1958), p. 120.

[40]That title is given on the title page to the sixth edition, published in New York by D. Appleton and Company, 1889. It is interesting that the modern additions omit the secondary title, which is included in full here, presumably because the implied racism of Mr. Darwin's theory is embarrassing in this enlightened age.
[41]W. Stanley Jevons, *The Coal Question: An Inquiry Concerning the Progress and a Probable Exhaustion of Our Coal Mines*, 2d ed. (London: Macmillan & Co, 1866), p. xxi.
[42]Eric Morris, "From Horse Power to Horsepower," *Access Magazine*, volume 30, Spring 2007, online at http://www.uctc.net/access/30/Access%2030%20-%2002%20-%20Horse%20Power.pdf, viewed 7/2/2008.
[43]Editorial, "Time to Banish the God of Growth," *New Scientist*, October 15, 2008, online at http://www.newscientist.com/article/mg20026782.500-time-to-banish-the-god-of-growth.html, viewed 6/23/2010.
[44]Jevons, *Coal Question*, p. 376.
[45]Eric Pianka, quoted from his presentation at the 109th meeting of the Texas Academy of Science March 2-4, 2006, transcribed by J. Richard Pearcey, online at http://www.pearceyreport.com/archives/2006/04/transcript_dr_d.php/, viewed 7/2/2009.
[46]Mark Steyn, "Doom If Saint Al Loses Carbs," *The Australian*, October 15, 2007, online at http://www.theaustralian.com.au/news/opinion/doom-if-saint-al-loses-carbs/story-e6frg6zo-1111114641484, viewed 6/23/2010.
[47]John Davis, quoted in Don Feder, "The Global Warming Suicide Cult," *FrontPageMag.com*, December 19, 2007, online at http://www.frontpagemag.com/Articles/Read.aspx?GUID=D9F17F62-10FF-4F53-8288-28353B567FDF, viewed 7/2/2009.
[48]D. M. Graber, "Mother Nature as a Hothouse Flower," *Los Angeles Times Book Review*, October 22, 1989, p. 10, online at http://articles.latimes.com/1989-10-22/books/bk-726_1_bill-mckibben, viewed 6/23/2010.
[49]James Lovelock, *Healing Gaia: Practical Medicine for the Planet* (New York: Harmony, 1991).
[50]Online at http://www.imdb.com/title/tt0133093/quotes, viewed 7/2/2009.
[51]Fleur Cowles and Prince Philip, *If I Were an Animal* (New York: William Morrow, 1987), foreword.
[52]Paul Ehrlich, *The Population Bomb* (New York: Ballantine Books, 1970), p. 166.
[53]Ehrlich, *Population Bomb*, p. 166.
[54]Online at www.earthday.gov, viewed 7:3:2009.
[55]Charles P. Alexander, "The Year In Nature," *Time Magazine*, December 31, 2000, online at http://www.time.com/time/magazine/article/0,9171,998892,00.html, viewed, 6/23/2010.
[56]Speech by Former Vice President Al Gore, New York University School of Law, September 18, 2006, online at http://www.nyu.edu/community/gore.html, viewed 7/3/2009.
[57]John Diaz, "An interview with Sen. Barack Obama," *San Francisco Chronicle*, January 22, 2008, online at http://www.sfgate.com/cgi-bin/blogs/sfgate/month?blogid=42&year=2008&month=01, viewed 11/1/2008.

[58]Mark Clayton, "Global Boom in Coal Power—and Emissions," *The Christian Science Monitor*, March 22, 2007, online at http://www.csmonitor.com/ 2007/ 0322/p01s04-wogi.html, viewed 7/9/2010.
[59]Ben Jackson, "Fatties Cause Global Warming," *The Sun*, April 20, 2009, online at http://www.thesun.co.uk/sol/homepage/news/article2387203.ece, viewed 7/9, 2010.
[60]"Global Warming: Now it Hits Brothels," *Metro.co.uk*, March 6, 2007, online at http://www.metro.co.uk/weird/article.html?in_article_id=39945&in_page_id=2, viewed 7/3/2009.
[61]"Researchers Find Global Warming Lengthens the Day," *AMS Newsletter*, 23:3 (March 2002).
[62]Bob Dow, "Veteran Loch Ness Monster Hunter Gives Up," *The Daily Record*, February 13, 2008, online at http://www.dailyrecord.co.uk/news/scottish-news/ 2008/02/13/veteran-loch-ness-monster-hunter-gives-up-86908-20317853/, viewed 7/9,2010.
[63]A. Pretia, G. Lentinic, and M. Maugeri , "Global Warming Possibly Linked to an Enhanced Risk of Suicide: Data from Italy, 1974–2003," *J. Affective Disorders* 102(1-3) (September 2007, 19-25, abstract online at http://www.ncbi.nlm.nih. gov/pubmed/17234275, viewed 7/9,2010.
[64]"Two-headed Olive Ridley Turtle Hatchling," World Wildlife Fund, November 29, 2005, online at http://www.panda.org/about_wwf/where_we_work/latin_ america_and_caribbean/country/c, viewed on 11/27/2007.
[65]Aaron Wildavsky, "No Risk Is the Highest Risk of All," in *Readings in Risk,* edited by Theodore S. Glickman and Michael Gough (London: RFF Press, 1990), p. 120.
[66]The Voluntary Human Extinction Movement (online at http://www.vhemt.org/, viewed 7/9/2010) proclaims that "Phasing out the human race by voluntarily ceasing to breed will allow Earth's biosphere to return to good health. Crowded conditions and resource shortages will improve as we become less dense."
[67]Gore was awarded the 2007 Nobel Peace Prize. The socialist Norwegian Parliament appoints the Norwegian Nobel Committee, which then selects the Laureate for the Peace Prize.
[68]Al Gore, *The Assault on Reason* (New York: Penguin, 2008). The book was for a time in first place on the New York Times bestsellers list.
[69]Gore starred in the movie *An Inconvenient Truth*, which won an Academy Award.
[70]Bryan Walsh, "Can We Save the World by 2015?" *Time*, December 1, 2007, online at http://www.time.com/time/health/article/0,8599,1689985,00.html, viewed 7/3/2009.
[71]Extracts of Gore's Nobel Prize for Peace Lecture on December 10, 2007, are online at http://nobelprize.org/nobel_prizes/peace, viewed 5/3/2008.
[72]Steven Mufson, "Power Plant Rejected Over Carbon Dioxide For First Time," *Washington Post*, October 19, 2007, p. A01: "The Kansas Department of Health and Environment yesterday became the first government agency in the United States to cite carbon dioxide emissions as the reason for rejecting an air permit for a proposed coal-fired electricity generating plant, saying that the greenhouse gas threatens public health and the environment." Online at http://www. washingtonpost.com/wp-dyn/content/article/2007/10/18/AR2007101802452. html, viewed 7/3/2009.

[73]The Liberal Democrat Party in Britain has committed itself to phasing out new petrol and diesel cars by 2040 as part of ambitious plans for a zero carbon Britain. "Lib Dems Plan to Scrap Petrol and Diesel Cars by 2040," *Edinburgh Evening News*, September 17, 2007, online at http://edinburghnews.scotsman.com/alternativeenergysources/Lib-Dems-plan-to-scrap.3327843.jp, viewed 7/9/2010.
[74]Rep. Jane Harman (D-CA) introduced legislation into the U.S. House of Representatives that would set target dates for certain types of light bulbs to be prohibited for sale in the United States. She said, "In this bill, we ban by 2012 the famously inefficient 100-watt incandescent bulb." Richard Simon, "Energy Bill Boosts Fuel Economy Standards," *Los Angeles Times*, December 19, 2007, online at http://www.latimes.com/news/nationworld/la-na-energy19dec19,0,1969731.story, viewed 7/9/2010.
[75]James M. Taylor, "Wood-burning Fireplaces May Be Banned in California," *Environment & Climate News*, February 2003, online at http://www.heartland.org/Article.cfm?artId=11433, viewed 7/3/2009.
[76]Geoffrey Lean, "A World Dying, But Can We Unite to Save It?" *The Independent online*, November 18, 2007, online at http://environment.independent.co.uk/climate_change/article3172144.ece, viewed 8/16/2008.
[77]Kathleen Norris, *Cloister Walk* (New York: Riverhead Books, 1996), p. 92.
[78]Al Gore, *Earth in the Balance*, rev. ed. (Boston: Houghton Mifflin Harcourt, 2000, p. 245.
[79]Al Gore, "The Moment of Truth," *Vanity Fair*, May 2006, online at http://www.vanityfair.com/politics/features/2006/05/gore200605, viewed 7/9/2010.
[80]Marcus Chown, "Has Observing the Universe Hastened Its End?" *New Scientist*, November 22, 2007, online at http://www.newscientist.com/article/mg19626313.800-has-observing-the-universe-hastened-its-end.html, viewed 7/9/2010.
[81]Sallie McFague, *The Body of God: An Ecological Theology* (Philadelphia: Fortress Press, 1993), pp. vii-viii.
[82]Extracts of Gore's Nobel Prize for Peace Lecture on December 10, 2007, are online at http://nobelprize.org/nobel_prizes/peace, viewed 5/3/2008.
[83]Steven Milloy, "Al Gore's Inconvenient Electric Bill," *Canada Free Press*, March 12, 2007 online at http://www.canadafreepress.com/2007/milloy031207.htm, viewed 7/9/2010.
[84]Online at http://TerraPass.
[85]Christopher Manes, *Green Rage: Radical Environmentalism and the Unmaking of Civilization* (Boston: Little, Brown, and Company, 1990), p. 34.
[86]Manes, *Green Rage*, p. 234.
[87]Roger S. Gottlieb, "Religion in an Age of Environmental Crisis," in *This Sacred Earth: Religion, Nature, Environment*, ed. Roger S. Gottlieb (New York: New York: Routledge, 2004), p. 13.
[88]John Calvin, *Commentary on Genesis* (Edinburgh: Banner of Truth Trust, 1984), p. 209.
[89]Online at http://www.vhemt.org/, viewed 7/3/2009.
[90]Natasha Courtney-Smith and Morag Turner, "Meet the Women Who Won't Have Babies—Because They're Not Eco-friendly," *Daily Mail*, November 21, 2007,

online at http://www.dailymail.co.uk/pages/live/femail/article.html?in_article_id=495495&in_page_id=1879, viewed 7/3/2009.

[91]Courtney-Smith and Turner, "Meet the Women."

[92]B. N. J. Walters, "Personal Carbon Trading: A Potential 'Stealth Intervention' for Obesity Reduction?" *Medical Journal of Australia*, Letters, 187:11:12 (December 3 and 17, 2007), 668.

[93]Katie McCabe, "Who Will Live, Who Will Die?" *Washingtonian* 21:11 (August 1986), p. 115.

[94]Quoted in Paul A. Clarke, *Dictionary of Ethics, Theology and Society* (New York: New York: Routledge, 1995), p. 263.

[95]Philip Longman, *The Empty Cradle: How Falling Birthrates Threaten World Prosperity And What To Do About It* (New York: New York: Basic Books, 2004), p. 167.

[96]John Foxe, *Foxe's Book of Martyrs* (Whitefish, MT: Kessinger Publishing, 2004), pp. 21-22.

[97]Max Oelschlaeger, "Caring for Creation," in *This Sacred Earth*, ed. Gottlieb, p. 539.

[98]Max Weber, *The Protestant Ethic and the Spirit of Capitalism* (New York: Charles Scribner's Sons, 1958).

[99]Herbert Schlossberg, *Idols for Destruction* (Nashville: Thomas Nelson, 1983), p. 72.

[100]Ralph Waldo Emerson, quoted in Michael Horton, ed., *The Agony of Deceit* (Chicago, Moody Press, 1990), p. 23.

[101]Benjamin F. Morris, *The Christian Life and Character of the Civil Institutions of the United States* (Philadelphia: George W. Childs; Cincinnati: Rickey & Carroll, 1864), p. 333.

[102]Herbert London, *America's Secular Challenge: The Rise of a New National Religion* (New York and London: Encounter Books, 2008).

[103]Quoted in John Robbins, "The Educational Establishment versus Civilization," *The Trinity Review* December 1997/January 1998, p. 11, online at http://www.trinityfoundation.org/PDF/154a-TheEducationalEstablishment.pdf, viewed 7/9/2010.

[104]Dan Gilbert, *Crucifying Christ in Our Colleges* (San Diego: Danielle Publishers, 1933).

[105]Jonathan Groner, "In 'Grutter v. Bollinger' Amicus Avalanche, One Brief Stood Out," *Legal Times*, July 2, 2003, online at http://www.law.com/jsp/law/international/LawArticleIntl.jsp?id=900005535864, viewed 7/9/2010.

[106]*A Nation at Risk: The Imperative for Educational Reform* (Washington: United States Government Printing Office, 1983), 1.

[107]Too many academics use their bully pulpit to indoctrinate students, passionately trying to lead them to become true believers in various left-wing political causes. In his book *Save the World on Your Own Time* (New York: Oxford University Press, 2008), Stanley Fish argues that academics should stick to "doing their jobs" and "introduce students to disciplinary materials and equip them with the necessary analytic skills" to engage in disciplined research.

[108]To express their rejection of "*pro Christo et ecclesia*," the new motto of Harvard excises Christ and His Church and reads simply, "*Veritas*."

[109]See, for example, Jeremiah 18:6-10 with Matthew 10:29-31 and 1 Kings 18:40-44.
[110]When Satan sought to tempt Him to lust after power, Jesus answered and said, "It is written, Man shall not live by bread alone, but by every word that proceeds from the mouth of God" (Matthew 4:4). He is Himself the eternal Word, but for our education honors the written Word, that we may know God's Word can be trusted.
[111]Quoted in James W. Sire, *Scripture Twisting: Twenty Ways the Cults Misread the Bible* (Downers Grove: InterVarsity Press, 1980), p. 7.
[112]Kenneth L. Woodward, "Finding God," *Newsweek*, February 7, 2000, online at http://www.newsweek.com/2000/02/06/finding-god.html, viewed 7/13/2010.
[113]David Remnick, "Profiles: The Wilderness Campaign: Al Gore Lives on a Street in Nashville," *The New Yorker*, September 13, 2004, online at http://www.newyorker.com/fact/content/?040913fa_fact, viewed 12/10/2005.
[114]Gore, *Earth in the Balance* (2000), p. 263.
[115]Gore, *Earth in the Balance* (2000), p. 260.
[116]Gore, *Earth in the Balance* (2000), p. 259.
[117]Stanley Fish, "Why We Can't All Just Get Along," *First Things*, 60 (Feb. 1996), 18-26, online at http://www.leaderu.com/ftissues/ft9602/articles/fish.html, viewed 7/13/2010.
[118]Blaine Harden, "The Greening of Evangelicals," *The Washington Post*, February 6, 2005, p. A01, online at http://www.washingtonpost.com/wp-dyn/articles/A1491-2005Feb5.html, viewed 7/13/2010.
[119]U.S. Religious Landscape Survey, "Religious Affiliation: Diverse and Dynamic," Pew Forum, February 2008, p. 5, online at http://religions.pewforum.org/pdf/report-religious-landscape-study-full.pdf, viewed 7/13/2010.
[120]Michael Weisskopf, "Energized by Pulpit or Passion, the Public Is Calling: `Gospel Grapevine' Displays Strength in Controversy Over Military Gay Ban," *Washington Post*, February 1, 1993, p. A01, online at http://dickstaub.com/links_view.php?record_id=3321, viewed 7/13/2010.
[121]Jesus warned that secular revolutionaries would attempt to fuse their concepts with Christian dialogue: "Beware of the false prophets, who come to you in sheep's clothing, but inwardly are ravening wolves. You will know them by their fruits" (Matthew 7:15–16).
[122]See, for example, Steven Rockefeller, *John Dewey: Religious Faith and Democratic Humanism* (New York: Columbia University Press, 1991).
[123]John Dewey, *A Common Faith* (New Haven, CT: Yale University Press, 1960), pp. 43, 51.
[124]Quoted in *Toward the Future of Reformed Theology: Tasks, Topics, Traditions*, edited by David Willis-Watkins, Michael Welker, Matthias Gockel, and David Willis (Grand Rapids: Eerdmans, 1998), p. 208.
[125]Gordon H. Clark, *Thales to Dewey* (Unicoi, TN: Trinity Foundation, 2000), pp. 377-382.
[126]Gordon H. Clark, *Three Types of Religious Philosophy*, (Unicoi, TN: Trinity Foundation, 1989), p. 114.
[127]Extracts of Gore's Nobel Prize for Peace Lecture on December 10, 2007, are online at http://nobelprize.org/nobel_prizes/peace, viewed 5/3/2008.
[128]Barna Group, "The God Gap? The Faith of Republicans and Democrats," March 5, 2007, online at http://www.barna.org/barna-update/article/13-culture/108-

the-god-gap-the-faith-of-republicans-and-democrats?q=evangelical, viewed on 11/11/2008.

[129]U.S. Religious Landscape Survey, "Religious Affiliation."

[130]Edward O. Wilson, *Naturalist* (Washington: Island Press, 2006), p. 368.

[131]Edward O. Wilson, *Consilience: The Unity of Knowledge* (New York: Vintage, 1999): "The human mind evolved to believe in the gods." Quoted in Natalie Angier, "The Bush Years; Confessions of a Lonely Atheist," *New York Times Magazine*, January 14, 2001, online at http://www.nytimes.com/2001/01/14/magazine/the-bush-years-confessions-of-a-lonely-atheist.html, viewed 7/13/2010.

[132]Edward O. Wilson, *The Creation: An Appeal to Save Life on Earth* (New York: Norton, 2006), p. 167.

[133]Fred Branfman, "Living in Shimmering Disequilibrium," *Salon Magazine*, April 22, 2000, online at http://www.salon.com/people/feature/2000/04/22/eowilson, viewed 11/15/ 2008.

[134]Wilson, *Consilience*, p. 80.

[135]Leary, Timothy, *Flashbacks, an Autobiography*, J.P. Tarcher, 1983, p. 91.

[136]Branfman, "Living in Shimmering Disequilibrium."

[137]J. Gresham Machen, *Christianity and Liberalism* (Grand Rapids: Eerdmans, 1923), p. 123.

[138]Gordon H. Clark, "The Limits and Uses of Science," in *Horizons of Science*, edited by Carl F. H. Henry (New York: Harper and Row, 1978), pp. 259, 260.

[139]Max Oelschlaeger, *Caring for Creation: An Ecumenical Approach to the Environmental Crisis* (New Haven, CT: Yale University Press, 1996), p. 201.

[140]Extracts of Gore's Nobel Prize for Peace Lecture on December 10, 2007, are online at http://nobelprize.org/nobel_prizes/peace, viewed 5/3/2008.

[141]Francis A. Schaeffer, *The God Who Is There* (Downers Grove, IL: Intervarsity Press, 1968).

[142]Francis A. Schaeffer, *The Church at the End of the 20th Century* (Wheaton: Crossway Books, 1970), pp. 123, 124.

[143]Strong was the Secretary-General of the first UN Conference on the Human Environment, in Stockholm, Sweden, held in 1972 (Earth Summit I). He has since organized the Rio Environmental Summit in 1992 (Earth Summit II) and was the Senior Advisor to the Secretary-General of the United Nations, Kofi Annan. He is founder of The Earth Charter Alliance, along with David Rockefeller, who penned it, and United Nations Environment Program (UNEP).

[144]It is unclear that Strong is a professing Baha'i, but his personal religious views neatly fit into that template. For example, Baha'is believe in the eventual unity and fusion of world religions and that nation states will wither away, to be replaced by a world government.

[145]Interview of Maurice Strong, Conversations with Harold Hudson Channer, aired on July 5, 2001, online at http://www.youtube.com/watch?v=4XueBqv_dpM&feature=PlayList&p=386FF8033B16FD48&playnext_from=PL&playnext=1&index=107, quotes at approximately 37 minutes.

[146]Crichton summarizes his view of modern environmentalism as "near-hysterical preoccupation with safety that's at best a waste of resources and a crimp on the human spirit, and at worst an invitation to totalitarianism." *State of Fear* (San Francisco: HarperCollins/Avon Books, 2004), p. 571.

[147]Maurice F. Strong, *Where on Earth Are We Going?* (Texere, 2001).
[1148]Michael Gilchrist, "Global Warming and Pagan Emptiness," interview of Cardinal George Pell, Catholic Education Resource Center, online at http://www.catholiceducation.org/articles/environment/print/en0015.htm, viewed 5/27/2009.
[149]George Marsden, *The Outrageous Idea of Christian Scholarship* (Oxford, UK: Oxford University Press, 1997), p. 81.
[150]Lynn White, Jr., "The Historic Roots of Our Ecologic Crisis," *Science*, 155 (March 10, 1967), pp. 1203-1207.
[151]Bill McKibben, *The End of Nature* (New York: Random House, 1989), pp. 74, 75.
[152]René Dubos, "Unity through Diversity," in *Who Speaks for Earth*, edited by Maurice F. Strong (New York: Norton, 1973), pp. 33-42.
[153]The World Council of Churches (WCC) is the international parent organization of the NCC. It bears a heavy burden in claiming to represent the ecumenical voices of well over half a billion Protestants in more than 110 countries, though not the voices of the over 1 billion Roman Catholics. Both the NCC and WCC have at various times advocated for the Green Dragon.
[154]Oelschlaeger, *Caring for Creation*, p. 204.
[155]George Dugan, "National Council of Churches Backs Artificial Birth Control," *New York Times*, February 24, 1961, p. 1.
[156]Steven Bouma-Prediger, *The Greening of Theology: The Ecological Models of Rosemary Radford Ruether, Joseph Stiller, and Jurgen Moltmann* (Atlanta: American Academy of Religion, 1995), pp. 61-91.
[157]Religious Coalition for Reproductive Choice, "Our Mission," online at http://www.rcrc.org/about/index.cfm, viewed 7/14/2010.
[158]John H. Cushman, "Religious Groups Mount a Campaign to Support Pact on Global Warming," *New York Times*, August 15, 1998, p. A10.
[159]National Council of Churches of Christ, "Eco-Justice Programs," online at http://www.nccecojustice.org/about/, viewed 7/14/2010.
[160]National Council of Churches, Eco Justice Program, "The Poverty of Global Climate Change," Presbyterian Church (USA), 2008, cited in *SciTech, Journal of the Presbyterian Association on Science, Technology, and the Christian Faith*, 17:2 (May 2008).
[161]HarperCollins Publishers promotional page for *The Green Bible*, online at http://www.harpercollins.com/books/Green-Bible-Harper-Bibles/?isbn=9780061627996, viewed 7/14/2010.
[162]Robert Booth Fowler, *The Greening of Protestant Thought* (Durham: University of North Carolina Press, 1995), p. 84.
[163]Amanda Little, "Cizik Matters: An Interview with Green Evangelical Leader Richard Cizik," *Grist.com*, online at http://www.grist.org/news/maindish/ 2005/10/05/cizik/, viewed 12/5/2008.
[164]See for example Lomborg, *Skeptical Environmentalist* and *Cool It: The Skeptical Environmentalist's Guide to Global Warming* (New York: Knopf, 2007) and Roy W. Spencer, *Climate Confusion: How Global Warming Hysteria Leads to Bad Science, Pandering Politicians and Misguided Policies that Hurt the Poor* (New York: Encounter Books, 2008) and *The Great Global Warming Blunder: How Mother Nature Fooled the World's Top Climate Scientists* (New York: Encounter Books, 2010).

[165]Mark D. Tooley, "Remaking Evangelicals in Ted Turner's Image?" *FrontPageMag.com*, May 14, 2009, online at http://www.frontpagemag.com/readArticle.aspx?ARTID=34866, viewed 8/8/2009.
[166]Linnie Marsh Wolfe, *Son of the Wilderness: The Life of John Muir*, 2d ed. (Madison: University of Wisconsin Press, 2003), p. 115.
[167]Sierra Club, *Faith in Action: Communities of Faith Bring Hope for the Planet* (San Francisco: Sierra Club, 2008), p. 3, online at http://www.sierraclub.org/ej/downloads/faithinactionreport2008.pdf, viewed 7/13/2010.
[168]Frank Pastore, "The National Council of Churches Should Have Died," *Townhall.Com*, March 25, 2007, online at http://townhall.com/columnists/FrankPastore/2007/03/25/the_national_council_of_churches_should_have_died, viewed 10/20/2008.
[169]Bron Taylor, "A Green Future for Religion?" *Futures*, 36 (2004), 991-1008.
[170]V. Kann Rasmussen Foundation, "We Have Special Interest in Greenwood and South Carolina," online at http://www.vkrf.org/vkrf_main.php?PID=CAT5, viewed 7/14/2010.
[171]See the organizations' websites at http://www.nrpe.org/index.html, http://www.creationcare.org/, and http://www.coejl.org/index.php, all viewed 7/14/2010.
[172]*Earthkeeping News*, A Newsletter of the North American Coalition for Christianity and Ecology (NACCE), Vol. 13, No. 3 (May:June 2004).
[173]Teresa Watanabe, "The Green Movement Is Getting Religion," *Los Angeles Times*, December 25, 1998, p. A-1.
[174]Michael L. Yoder, "Mennonites, Economics, and the Care of Creation," in *Creation and the Environment: An Anabaptist Perspective on a Sustainable World*, edited by Calvin Wall Redekop (Baltimore: Johns Hopkins University Press, 2000), p. 88.
[175]Christian Schools that teach Au Sable courses are listed at http://www.ausable.org/ap.colleges.cfm#participating, viewed 12/7/2008.
[176]Jeffrey Golliher and William Bryant Logan, ed., *Crisis and the Renewal of Creation: World and Church in the Age of Ecology* (New York: Continuum International Publishing Group, 1996).
[177]Ari L. Goldman, "Ecumenically Environmental," *New York Times*, Religion Notes, October 9, 1993, p. 11.
[178]Quoted in Michael Barkey, "A Green Reformation? Saving Faith from the Environmentalists," Acton Institute, March 23, 2000, online at http://www.acton.org/commentary/commentary_12.php, viewed 7/14/2010.
[179]George Orwell, *Animal Farm* (Fairfield, IA: 1st World Publishing, 2005), p. 11.
[180]Arne Næss, *Ecology, Community and Lifestyle: Outline of Ecosophy*, translated and revised by David Rothenberg (Cambridge: Cambridge University Press, 1989), p. 28.
[181]Aldo Leopold, *Sand County Almanac* (New York: Ballantine Books, 1986).
[182]John Muir, *The Eight Wilderness Discovery Books* (Kennoway, Fife, UK: Diadem Books, 1992), pp. 160-161.
[183]See, for example, Eccy De Jonge, *Spinoza and Deep Ecology: Challenging Traditional Approaches to Environmentalism* (Surrey, UK: Ashgate New Critical Thinking in Philosophy), 2004, p. 35.

[184]Manes, *Green Rage*, p. 142.
[185]Pierre Teilhard de Chardin, *The Phenomenon of Man* (New York, Harper and Brothers, 1959).
[186]Rudolf Krügel, *Der Begriff des Volksgeistes in Ernst Moritz Arndts Geschichtsanschauung* (Langensalza, 1914), quoted in Peter Staudenmaier, "Fascist Ecology: The 'Green Wing' of the Nazi Party and its Historical Antecedents," in *Ecofascism: Lessons from the German Experience*, by Janet Biehl and Peter Staudenmaier (Edinburgh: AK Press, 1995), p. 6.
[187]Cited in Fowler, *The Greening of Protestant Thought*, p. 100.
[188]Hubert L. Dreyfus, "On the Ordering of Things: Being and Power in Heidegger and Foucault," *Michel Foucault, Philosophe* (Paris: Le Seuil, 1989).
[189]Naess, *Ecology, Community and Lifestyle*, p. 163.
[190]Gore, *Earth in the Balance* (2000), p. 215.
[191]Roger S. Gottlieb, ed., *This Sacred Earth*, p. 11.
[192]Sallie McFague, *Models of God: Theology for an Ecological, Nuclear Age* (Philadelphia: Fortress, 1987), pp. 29-31, 60, and 23.
[193]De Jonge, *Spinoza and Deep Ecology*.
[194]"Save the Planet? That's So Cool...," *The Independent*, May 9, 2005, online at http://www.independent.co.uk/environment/save-the-planet-thats-so-cool-490072.html, viewed 7/14/2010.
[195]Online at http://www.ema-online.org/ema_video_playful.php, viewed 11/26/2008.
[196]Extracts of Gore's Nobel Prize for Peace Lecture on December 10, 2007, are online at http://nobelprize.org/nobel_prizes/peace, viewed 5/3/2008.
[197]Cited in Joanne Hollows, *Feminism, Femininity and Popular Culture* (Manchester, UK: Manchester University Press, 2000), p. 199.
[198]Oprah Winfrey in Internet interview with Eckhart Tolle, March 2008, online at http://www.youtube.com/watch?v=JW4LLwkgmqA, viewed 10/10/2008.
[199]Machen, *Christianity and Liberalism*, pp. 77, 78.
[200]Janet Biehl and Peter Staudenmaier, *Ecofascism: Lessons from the German Experience* (Edinburgh: AK Press, 2001), p. 50.
[201]John Passmore, *Man's Responsibility for Nature: Ecological Problems and Western Traditions* (New York: Scribner's, 1974), pp. 39-40.
[202]David Spangler, "Imagination, Gaia, and the Sacredness of the Earth," in *This Sacred Earth*, ed. Gottlieb, p. 618.
[203]Rudyard Kipling, *Rudyard Kipling's Verse* (New York: Doubleday Page, 1919), p. 770.
[204]In the absence of other radiation, the cosmic microwave background temperature is about 2.7 Kelvin, or -454.5°F.
[205]Carl Sagan, *Cosmos* (New York: Random House, 1980), p. 140.
[206]L. Baum and P. H. Frampton, "Turnaround in Cyclic Cosmology," *Phys. Rev. Lett.* 98, 071301 (2007), online at http://www.physics.unc.edu/~frampton/TURN.pdf, viewed 7/14/2010.
[207]Rosemary Radford Ruether, *Gaia and God: An Ecofeminist Theology of Earth Healing* (San Francisco: Harper Collins, 1992), p. 15.
[208]McFague, *Body of God*, pp. ix-x.

[209]Karl Popper, *Conjectures and Refutations* (New York: Routledge Classics, 2006), p. 186.
[210]McFague, *Models of God*, p. xi.
[211]E. Calvin Beisner, "Deep Ecology, Neo-Paganism, and the Irrationalism of Global Warming Hysteria," CWIPPThink 2008, Escondido, California, January 21-24, 2008, online at http://www.ecalvinbeisner.com/freearticles/DeepEcol&GlobalWarming.pdf, viewed 2/14/2009; condensed and reprinted in *On Global Wizardry: Techniques of Pagan Spirituality and a Christian Response*, edited by Peter Jones (Escondido, CA: Main Entry Editions, 2010), 171-185.
[212]Franz Rosenzweig, *The Star of Redemption* (Madison: University of Wisconsin Press, 2005), p. 11.
[213]C. S. Lewis, *The Last Battle* (London: Puffin Books, 1956), p. 132.
[214]Jeffrey Kripal, *Esalen: America and the Religion of No Religion* (Chicago: University of Chicago Press, 2007), pp. 298-314.
[215]Brian Swimme, *The Universe Is a Green Dragon: A Cosmic Creation Story* (Rochester, VT: Bear & Company, 1984).
[216]Gary Zukav, *The Dancing Wu Li Masters: An Overview of the New Physics* (New York: Bantam, 1984).
[217]John Milton, *Paradise Lost* (New York: Penguin Classics, 2003), p. 28.
[218]Alan Sokal and Jean Bricmont, *Fashionable Nonsense: Postmodern Intellectuals' Abuse of Science* (New York: Macmillan:Picador, 1999).
[219] Dewey, *Common Faith*, p. 43.
[220]Gore, *Earth in the Balance* (2000), p. 259.
[221]Mary Evelyn Tucker, "The Ecological Spirituality of Pierre Teilhard de Chardin," *Spiritus: A Journal of Christian Spirituality*, 7:1 (Spring 2007), pp. 1-19.
[222]Erazim Kohák, *The Green Halo: A Bird's-Eye View of Ecological Ethics* (Chicago: Open Court, 1999), p. 25.
[223]For example, consider John 14:6 with John 10:7–10 and Galatians 1:6–9.
[224]Deuteronomy 6:4; 32:39; 1 Timothy 2:5. Scripture says: "Great is our Lord, and mighty in power: His understanding is infinite" (Psalm 147:5).
[225]Schaeffer, *Church at the End of the Twentieth Century*, p. 112.
[226]Carl Sagan and Ann Druyan, *The Demon-Haunted World: Science as a Candle in the Dark* (New York: Ballantine Books, 1997).
[227]Sagan, *Cosmos*, p. 4.
[228]Revelation 4:11.
[229]Cited in Arthur Pink, *Gleanings from the Scripture: Man's Total Depravity* (Chicago: Moody Press, 1977), p. 11.
[230]Calvin, *Commentary on Genesis*, p. 92.
[231]Peter Singer, *Practical Ethics*, 2d ed. (Cambridge: Cambridge University Press, 1999), p. 331.
[232]Robert Shaw, *An Exposition of the Confession of Faith of the Westminster Assembly of Divines* (Philadelphia: Presbyterian Board of Publication, 1845), p. 80.
[233]Jonathan Edwards, *The Blank Bible*, in *The Works of Jonathan Edwards*, 26 volumes, multiple editors (New Haven, CT: Yale University Press, 2006), vol. 24, p. 135.
[234]Cotton Mather, *The Christian Philosopher*, edited by Winton Solberg (Champaign, IL: University of Illinois Press, 2000), p. 236.

[235]Albert Barnes, *Barnes' Notes on the New Testament* (Grand Rapids: Kregel Classics, 1962), p. 1451.
[236]Vanessa Lemm, *Nietzsche's Animal Philosophy: Culture, Politics, and the Animality of the Human Being* (New York: Fordham University Press, 2009).
[237]Singer, *Practical Ethics*, p. 272.
[238]William Shakespeare, *Hamlet* (Oxford: Oxford University Press, 2002), p. 52.
[239]Jacques Derrida, "And Say the Animal Responded?" in *Zoontologies: The Question of the Animal*, translated by David Willis, edited by Cary Wolfe (Minneapolis: University of Minnesota Press. 2003), p. 137.
[240]Blaise Pascal, "The Greatness of Man's Dignity," in *The Mind on Fire*, edited by James M. Houston (Vancouver: Regent College Publishing, 1989), p. 82.
[241]John Calvin, *Institutes of the Christian Religion*, translated by Ford Lewis Battles, edited by John T. McNeill (Philadelphia: Westminster John Knox Press, 1960), book III, chapter xi, p. 742.
[242]Gore, *Assault on Reason*, p. 56.
[243]Friedrich Nietzsche, *Thus Spake Zarathustra* (Whitefish, MT: Kessinger Publishing, 2004), p. 93.
[244]G. K. Chesterton, *Orthodoxy* (Vancouver: Regent College Publishing, 2004), p. 20.
[245]James Henley Thornwell, *The Collected Writings of James Henley Thornwell*, edited by John B. Adger (Richmond, VA: Presbyterian Board of Publication, 1870), vol. 1, p. 237.
[246]Carl Friedrich Keil and Franz Delitzsch, *Biblical Commentary on the Old Testament*, in *Clark's Foreign Theological Library*, (Edinburgh: T. & T. Clark, 1866), vol. 2, p. 99.
[247]Blaise Pascal, "The Greatness of Man's Dignity," in *The Mind on Fire*, ed. Houston, p. 88.
[248]Edwards, *Blank Bible*, in *Works*, vol. 24, p. 136.
[249]Gore, *Earth in the Balance* (2000), p. 258.
[250]Gore, *Earth in the Balance* (2000), p. 256.
[251]Quoted on *Arts Initiative at Columbia University* postcard, 2009.
[252]Robert Jamieson, A. R. Fausset, and D. Brown, *A Commentary Critical and Explanatory on the Old and New Testaments*, (Glasgow: William Collins, 1863), vol. 2, p. 299.
[253]Jonathan Edwards, "Miscellanies," 1246, "The Saints Higher in Glory than the Angels," in *Works*, edited by Douglas A. Sweeney (New Haven, CT: Yale University Press, 2004), vol. 23, p. 182.
[254]Blaise Pascal, "Christian Morality," in *Mind on Fire*, ed. Houston, p. 219.
[255]Peter Singer, "Killing Babies Isn't Always Wrong," *The Spectator*, 16 September 1995.
[256]Ed Douglas, "The Mountain of Mountains," *The Observer*, August 31, 2008, online at http://www.guardian.co.uk/sport/2008/aug/31/9, viewed 12/7/2008.
[257]Quoted in Jonathan Waterman, *The Quotable Climber* (Guilford, CT: Lyons Press, 2002), p. 216.
[258]Lindsay Jones, Lindsay, ed., *Encyclopedia of Religion* (Farmington Hills, MI: Thomson Gale, 2005), vol. 4, p. 1119.
[259]Manes, *Green Rage*, p. 141.

[260]Matthew Fox, *Original Blessing: A Primer in Creation Spirituality* (Santa Fe, NM: Bear and Company, 1983), p. 46.
[261]Manes, *Green Rage*, p. 142.
[262]Friedrich Nietzsche, *Twilight of the Idols and the Antichrist* (Mineola, NY: Dover Publications, 2004), p. 85.
[263]Naess, *Ecology, Community and Lifestyle*, p. 168.
[264]Quoted in *Mensch & Energie*, April 1983.
[265]Cited in Ara Paul Barsam, *Reverence for Life: Albert Schweitzer's Great Contribution to Ethical Thought* (Oxford: Oxford University Press, 2008), pp. 28, 29.
[266]Victor Lasky, *The Ugly Russian* (New York: Pocket Books, 1966), p. 97.
[267]David W. Ehrenfeld, *The Arrogance of Humanism* (Oxford: Oxford University Press, 1981).
[268]Wade, *Before the Dawn*.
[269]Spengler (David P. Goldman), "Harry Potter and the Decline of the West," *The Asia Times*, 20 July 2005, online at http://www.atimes.com/atimes/Front_Page/GG20Aa01.html, viewed 2/13/2009.
[270]Carl Sagan, *Cosmos* (New York: Random House, 1980), p. 1.
[271]David Jennings, *Jewish Antiquities: or a course of lectures on the three first books of Godwin's Moses and Aaron. To which is annexed, a dissertation on the Hebrew language* (William Bayne, and Sons, 1825), pp. 177-181.
[272]Christopher McIntosh, "The Pagan Revival and its Prospects," *Futures*, vol. 36, no. 9 (November 2004), pp. 1037-1041.
[273]Ben Child, "James Cameron Hits Back at Man Who Said He Was 'Running for Antichrist'," *The Guardian*, March 24, 2010, online at http://www.guardian.co.uk/film/2010/mar/24/james-cameron-glenn-beck-asshole, viewed 3/24/2010.
[274]Charles Spurgeon, *John Ploughman's talk: or Plain Advice for Plain People* (New York: Robert Carter & Bros., 1884), p. 60.
[275]Carl Sagan, "Preserving and Cherishing the Earth: An Appeal for Joint Commitment in Science and Religion," *Am. J. Physics*, vol. 58 no. 7 (July 1990), pp. 615-617.
[276]Schlossberg, *Idols for Destruction*.
[277]See http://www.earthcharter.org for the entire text, from which these quotes are drawn.
[278]Edward K. Kaplan, "Reverence and Responsibility: Abraham Joshua Heschel on Nature and the Self," Conference on Judaism and the Natural World, February 22-24, 1998, Harvard University, Center for the Study of World Religions, p. 12.
[279]Donald Moore, *The Human and the Holy: The Spirituality of Abraham Joshua Heschel* (New York: Fordham University Press, 1989), p. 93.
[280]David Spangler, "Imagination, Gaia, and the Sacredness of the Earth," in *This Sacred Earth*, ed. Gottlieb, p. 616.
[281]Ruether, *Gaia and God*, p. 48.
[282]Au Sable Institute, online at http://www.ausable.org/au.ausableidea.cfm and http://www.grist.org/news/maindish/2006/10/17/dewitt/, viewed 3/17/2009.
[283]Calvin DeWitt, in *This Sacred Earth*, ed. Gottlieb, p. i.
[284]Quoted in Fowler, *Greening of Protestant Thought*, p. 40.
[285]Fowler, *Greening of Protestant Thought*, p. 43.

[286]Brian J. Walsh, Marianne B. Karsh, and Nik Ansell, "Trees, Forestry, and the Responsiveness of Creation," in *This Sacred Earth*, ed. Gottlieb, p. 427.
[287]John Fox, *A History of the Lives, Sufferings, and Triumphant Deaths of the Primitive as well as the Protestant Martyrs from the Commencement of Christianity to the Latest Periods of Pagan and Popish Persecution*, Philadelphia, James M. Campbell, 1845, p. 11.
[288]Deborah Lipp, *The Elements of Ritual: Air, Fire, Water & Earth in the Wiccan Circle* (Woodbury, MN: Llewellyn Publications, 2003), p. 156.
[289]Üstün Bilgen-Reinart, *Porcelain Moon and Pomegranates: A Woman's Trek Through Turkey* (Toronto, ON: Dundurn Press, 2007), p. 71. "The earth mother was called many different names. In Anatolia she was Hepat, Kubaba, Kubele, Matar, Cybele, Agdistic, Leto, and Ephesian Artemis. The Sumerians called her Ma, Marienna, Inanna. Babylonians knew her as Ishtar, Astarte, Ashtoreth. In Egypt her name was Isis. In Crete she was Gaia or Rhea."
[290]Richard Dawkins, *The God Delusion* (Boston: Houghton Mifflin, 2006), p. 31.
[291]Robert L. Dabney, *The Sensualistic Philosophy of the Nineteenth Century Considered* (Anson D. F. Randolf & Company, 1887), p. 350.
[292]Dawkins, *God Delusion*, p. 14.
[293]Gore, *Earth in the Balance* (2000), p. 265.
[294]Richard Dawkins, *A Devil's Chaplain: Reflections on Hope, Lies, Science, and Love* (Boston: Mariner Books, 2004), p. 22.
[295]Dawkins, *God Delusion*, p. 5.
[296]Carl Sagan, *Cosmos* (DVD), directed by Adrian Malone (Los Angeles: Cosmos Studios, 2002).
[297]McKibben, *End of Nature*, p. 73.
[298]Sagan, *Cosmos* (DVD).
[299]Quoted in *The National Review*, 11 June 1982, p. 677.
[300]P. D. Ryan, *Buddhism and the Natural World: Towards a Meaningful Myth* (Cambridge, UK: Windhorse Publications, 2004), p. 43.
[301]Kip S. Thorne, "Spacetime Warps and the Quantum World," in *The Future of Spacetime*, edited by Alan Lightman (New York: W. W. Norton & Company, 2003), p. 148.
[302]Ken W. Campbell, *A Startrek to Eternity* (Parker, CO: Outskirts Press, 2007), pp. 63-65.
[303]Cotton Mather, *The Christian Philosopher*, edited by Winton Solberg (Champaign: University of Illinois Press, 2000), p. 306.
[304]Frederick S. Leahy, *Satan Cast Out: A Study in Biblical Demonology* (Carlisle, PA: Banner of Truth Trust, 1990), p. 13.
[305]Spangler, "Imagination, Gaia, and the Sacredness of the Earth," in *This Sacred Earth*, ed. Gottlieb, pp. 613, 614.
[306]J. G. Murphy, *A Critical and Exegetical Commentary on the Book of Genesis* (Boston: Draper and Halliday, 1867), p. 106.
[307]George Grant, *Killer Angel: A Biography of Planned Parenthood's Founder Margaret Sanger* (Franklin, TN: Ars Vitae Press and The Reformer Library, 1995).
[308]Margaret Sanger, *Woman and the New Race* (New York: Brentano's Publishers, 1920), p. 45.

[309]Y. J. Pendleton et al., "Voyages Through Time: Everything Evolves," American Geophysical Union Fall Meeting, San Francisco, *EOS. Trans. AGU*, 82(47), Fall Meet. Suppl. Abstract ED22A-04, 2001.
[310]Calvin, *Commentary on Genesis*, p. 111.
[311]Gordon H. Clark, *The Biblical Doctrine of Man* (Unicoi, TN: Trinity Foundation, 1992), p. 6.
[312]Blaise Pascal, in *Mind on Fire*, ed. Houston, p. 105.
[313]Graham Chapman, *Monty Python & Holy Grail* (London: Methuen Publishing Ltd, 2002).
[314]Calvin, *Commentary on Genesis*, p. 165.
[315]Edwards, in *Works*, vol. 6, p. 374.
[316]Dabney, *Sensualistic Philosophy of the Nineteenth Century*, p. 185.
[317]Pascal, in *Mind on Fire*, ed. Houston, p. 117.
[318]Spengler (David P. Goldman), "Sufism, Sodomy and Satan," *Asia Times Online*, August 12, 2008, online at http://www.atimes.com/atimes/Middle_East/JH12Ak03.html, viewed 7/17/2008.
[319]C. S. Lewis, *Prince Caspian* (New York: Harper Collins, 1994), p. 128.
[320]Ehrlich, *Population Bomb*, p. 15.
[321]Richard Dawkins, *The Selfish Gene* (Oxford: Oxford University Press, 1976).
[322]E. O. James, *Origins of Sacrifice* (Whitefish, MT: Kessinger Publishing, 2003).
[323]Binita Mehta, *Widows, Pariahs, and Bayaderes: India as Spectacle* (Lewisburg, PA: Bucknell University Press, 2002), p. 74.
[324]Ehrlich, *Population Bomb*, p. 156.
[325]Ehrlich, *Population Bomb*, p. 170.
[326]Ehrlich, *Population Bomb*, pp. 170-171.
[327]White, "Historic Roots of Our Ecologic Crisis,", pp. 1203-1207.
[328]Paul R. Ehrlich, *Human Natures: Genes, Cultures, and the Human Prospect* (Washington: Island Press, 2000), p. 261.
[329]Shepard Krech III, *The Ecological Indian: Myth and History* (New York: W. W. Norton & Company, 2000), pp. 123-150.
[330]George Washington to Timothy Pickering, July 1, 1796, in *The Writings of George Washington*, edited by John C. Fitzpatrick (Washington: U.S. Government Printing Office, 1931-1944), vol. 35, p. 112.
[331]Jonathan Edwards, *Life and Diary of David Brainerd* (New York: Cosimo Classics, 2007), pp. 214, 215.
[332]Wendell Berry, "Christianity and the Survival of Creation," in *Seeing God Everywhere: Essays on Nature and the Sacred*, edited by Barry McDonald (Bloomington, IN: World Wisdom, 2004), p. 53.
[333]Roderick F. Nash, *The Rights of Nature* (Madison: University of Wisconsin Press, 1989).
[334]White, "Historic Roots of Our Ecologic Crisis," pp. 1203-1207.
[335]Spangler, "Imagination, Gaia, and the Sacredness of the Earth," in *This Sacred Earth*, ed. Gottlieb, p. 616.
[336]Rashad Khalifa, ed., *Quran: The Final Testament* (Universal Unity, 2001), pp. 99, 100.
[337]Kohák, *Green Halo*, p. 18.

[338]Manes, *Green Rage*, p. 241.
[339]Presbyterian Eco-Justice Task Force Committee on Social Witness Policy, *Keeping and Healing the Creation* (Louisville, KY: Presbyterian Church USA, 1989), p. 2.
[340]Charles Darwin, quoted in Carl N. Degler, *In Search of Human Nature: The Decline and Revival of Darwinism in American Social Thought* (Oxford: Oxford University Press, 1993), p. 7.
[341]Wendell Berry, "Christianity and the Survival of Creation, " in *Seeing God Everywhere: Essays on Nature and the Sacred*, edited by Barry McDonald (Bloomington, IN: World Wisdom, 2004), p. 61.
[342]Christopher Monckton, "Great Is Truth, and Mighty Above All Things," keynote address to the 2009 International Conference on Climate Change, New York, 12 March 2009, online at http://www.heartland.org/full/24881/Great_Is_Truth_and_Mighty_Above_All_Things.html, viewed 7/15/2010.
[343]Indur M. Goklany, "Now for the Good News: Mankind has Never Been Healthier, Wealthier or Freer. Surprised?" *Reason*, March 23, 2007, online at http://reason.com/archives/2007/03/23/now-for-the-good-news, viewed 7/15/2010.
[344]Spengler (David P. Goldman), "Why Nations Die," *Asia Times Online*, August 16, 2005, online at http://www.atimes.com/atimes/Front_Page/GH16Aa02.html, viewed 7/25/2009. The precise quip Goldman attributes to Chesterton doesn't appear to be Chesterton's own words, but it neatly summarizes some of his thought.
[345]Naess, *Ecology, Community and Lifestyle*, p. 168.
[346]Wendell Berry, "Christianity and the Survival of Creation," in *Seeing God Everywhere*, ed. McDonald, p. 58.
[347]Alan Rodgman and Thomas A. Perfetti, *The Chemical Components of Tobacco and Tobacco Smoke* (London and New York: CRC Press, 2008), p. 933.
[348]Edwards, "Miscellanies," 833-1152, in *Works*, vol. 20, p. 81.
[349]Singer, *Practical Ethics*, p. 274.
[350]Peter Singer, *The President of Good and Evil: Questioning the Ethics of George W. Bush* (Essex, UK: Granta Books, 2004), also Peter Singer, "Quick on the Draw," *The Age*, April 17, 2004, online at http://www.theage.com.au/articles/2004/04/14/1081838786650.html?from=storyrhs, viewed 7/15/2010.
[351]Adolf, *Mein Kampf*, translated by James Murphy (London: Hurst and Blackett, 1939), p. 162.
[352]Biehl and Staudenmaier, *Ecofascism*, pp. 14-17.
[353]Colin Spencer, *The Heretic's Feast: A History of Vegetarianism* (Hanover, NH: University Press of New England, 1995), pp. 304-309.
[354]R. J. Rushdoony, *Roots of Reconstruction* (Vallecito, CA: Ross House Books, 1991), p. 412.
[355]Ezra Stiles, "The United States Elevated to Glory and Honour," in *The Pulpit of the American Revolution*, edited by John Wingate Thornton (Boston: D. Lothrop and Co., 1876), p. 404.
[356]Edwards, "Miscellanies," 833-1152, in *Works*, vol. 20, p. 103.
[357]Edwards, "Miscellanies," 833-1152, in *Works*, vol. 20, p. 521.
[358]John Calvin, *Calvin's Wisdom*, edited by Graham Miller (Carlisle, PA: Banner of Truth Trust, 1992), p. 185.

[359]Stephen Charnock, *Discourses Upon the Existence and Attributes of God* (New York: Robert Carter & Brothers, 1874), vol. 2, p. 373.
[360]Calvin, *Calvin's Wisdom*, ed. Miller, p. 184.
[361]Jonathan Edwards, "Sermons and Discourses, 1739-1742," in *Works*, vol. 22, p. 410.
[362]Cotton Mather, *The Christian Philosopher* (Champaign, IL: University of Illinois Press, 2000), p. 87.
[363]Mather, *Christian* Philosopher, p. 237.
[364]Mather, *Christian Philosopher*, p. 239.
[365]Calvin, *Calvin's Wisdom*, ed. Miller, p. 186.
[366]Simon, *Ultimate Resource 2*, p. xxxiv.
[367]Calvin De Witt, on the back cover of *This Sacred Earth*, ed. Gottlieb; Tri Robinson, *Saving God's Green Earth: Rediscovering the Church's Responsibility to Environmental Stewardship* (Norcross, GA: Ampelon, 2006).
[368]Quoted in Gavin Finefield, "I Can't Hear You: A Sermon on Earth Day Sunday 2008," *Journal of the Presbyterian Association on Science, Technology, and the Christian Faith*, Vol. 17, No. 2, May 2008, p. 9.
[369]Nicholas Dawidoff, "The Civil Heretic," *New York Times*, 29 March 2009, p. MM32.
[370]Paul Krugman, "Betraying the Planet," *New York Times*, June 29, 2009, p. A21, online June 28, 2009, at http://www.nytimes.com/2009/06/29/opinion/29krugman.html, viewed 7/15/2010; Marlo Lewis, "Ignorance Is Strength, Dissent Is Treason: Robert Kennedy Goes Orwell," *National Review*, July 16, 2007, online at http://article.nationalreview.com/321760/ignorance-is-strength-dissent-is-treason/marlo-lewis, viewed 7/15/2010.
[371]Ed Pilkington, "Put Oil Firm Chiefs on Trial, Says Leading Climate Change Scientist," *The Guardian*, 23 June 2008, online at http://www.guardian.co.uk/environment/2008/jun/23/fossilfuels.climatechange, viewed 7/15/2010.
[372]Kohák, *Green Halo*, p. 18.
[373]Václav Klaus, *Blue Planet in Green Shackles: What Is Endangered: Climate or Freedom?* (Washington: Competitive Enterprise Institute, 2008), p. 2.
[374]Edward N. Lorenz, *The Essence of Chaos* (Seattle: University of Washington Press,1996).
[375]Al Gore, *An Inconvenient Truth: The Planetary Emergency of Global Warming and What We Can Do About It* (New York: Rodale Books, 2006).
[376]Patrick Moore, "Why I left Greenpeace," *Wall Street Journal*, April 22, 2008, p. A23, online at http://online.wsj.com/article/SB120882720657033391.html (subscription required), viewed 7/15/2010.
[377]John M. Broder, "E.P.A. Clears Way for Greenhouse Gas Rules," *New York Times*, April 18, 2009, p. A15, online at http://www.nytimes.com/2009/04/18/science/earth/18endanger.html, viewed 7/15/2010.
[378]Howard Witt, "Canada's Forests, Once Huge Help on Greenhouse Gases, Now Contribute to Climate Change," *Chicago Tribune*, January 2, 2009, online at http://www.chicagotribune.com/news/nationworld/chi-canada-trees_wittjan02,0,539661.story, viewed 7/15/2010.
[379]Gore, *Earth in the Balance* (2000), p. 245.
[380]Ehrlich, *Population Bomb*, p. 148.

[381]Quoted in Finefield, "I Can't Hear You," p. 9.
[382]McKibben, *End of Nature*, pp. 58, 59.
[383]Fowler, *Greening of Protestant Thought*.
[384]Paul R. Ehrlich, and Anne M. Ehrlich, *One with Nineveh: Politics, Consumption, and the Human Future* (Washington: Island Press, 2004), p. 47.
[385]E. O. Wilson, *Biophilia* (Cambridge, MA: Harvard University Press, 1984).
[386]Elisabet Sahtouris, "The Conscious Universe," in *When Worlds Converge: What Science and Religion Tell Us about the Story of the Universe and Our Place in It*, edited by Clifford N. Matthews, Mary Evelyn Tucker, Philip J. Hefner (Chicago: Open Court, 2001), pp. 64-69.
[387]Henning Steinfeld, et al., *Livestock's Long Shadow—Environmental Issues and Options* (Geneva: United Nations Food & Agricultural Organization, 2006).
[388]Joanna Pearlstein, "Surprise! Conventional Agriculture Can Be Easier on the Planet," *Wired Magazine*, 19 May 2008, online at http://www.wired.com/science/planetearth/magazine/16-06/ff_heresies_03organics, viewed 7/15/2010.
[389]Jerry A. Coyne, *Why Evolution Is True* (New York: Viking Adult, 2009).
[390]Gary North, *The Dominion Covenant: Genesis*, rev. ed. (Tyler, TX: Institute for Christian Economics, 1987), p. 124.
[391]Ehrlich and Ehrlich, *One with Nineveh*, p. 24.
[392]Tenenbaum, David J., International Health: North Korean Catastrophe, *Environ. Health Perspect.*, vol. 113(1), January 2005, A26.
[393]Kang, Chol-Hwan and Pierre Rigoulot, *The Aquariums of Pyongyang*, New York: Basic Books, 2001, p. 104.
[394]Intergovernmental Panel on Climate Change (IPCC), *Climate Change 2007 - Mitigation of Climate Change: Working Group III contribution to the Fourth Assessment Report of the IPCC*, Cambridge University Press, 2007, pp. 707, 708.
[395]Goklany, Indur M., *The Precautionary Principle: A Critical Appraisal*, Cato Institute, 2001, p. 24.
[396]DeWitt, Calvin B., *Respecting Creation's Integrity: Biblical Principles for Environmental Responsibility*, p. 20.
[397]DeWitt, Calvin B., Preparing the Way for Action [Electronic version]. *Perspectives in Science and Christian Faith* 46, 1994, p. 82.
[398]Loscalzo, Craig A., *Preaching sermons that connect: Effective communication through identification*, InterVarsity Press, 2000.
[399]Goklany, Indur M., Richer is Cleaner: Long-term Trends in Global Air Quality, in (ed.) Ronald Bailey, *The True State of the Planet*, Free Press, 1995, esp. pp. 342-343.
[400]Clarke, Adam, *Commentary on the Holy Bible*, Volume 2, John J. Harrod, 1836, p. 440.
[401]Naess, *Ecology, Community and Lifestyle*, p. 29.
[402]Lee, Stan, Jack Kirby, and Steve Ditko, *Amazing Fantasy: Spider Man, # 15*, August 1962.
[403]Carson, Rachel, *Silent Spring*, Mariner Books, 2000, p. 246.
[404]Sidley, Pat, Malaria epidemic expected in Mozambique, *British Medical Journal*, Vol. 320 (11), 11 March 2000, p. 669.
[405]Lockitch, Keith, It's Time to Silence Silent Spring, *Capitalism Magazine*, 2 September 2007, online at http://www.capmag.com/article.asp?ID=5018i, viewed 11/11/2008.

[406]Heidegger, M., *Existence and Being*. Maurois Press, 2007, p. 157.
[407]Wilson, *Naturalist*, p. 87.
[408]Kohák, *Green Halo*, p. 42.
[409]McFague, Sallie, *Body of God*, p. 8.
[410]Matthew Fox, *Original Blessing: A Primer in Creation Spirituality Presented in Four Paths, Twenty-Six Themes, and Two Questions* (New York: Tardier, 2000), p. 231.
[411]Singer, *Practical Ethics*, p. 287.
[412]Rebecca English, "The Prince of Hypocrites: Charles Embarks on 16,000 Mile 'Green' Crusade ... Aboard a Private Jet," *The Daily Mail*, February 13, 2009, 3online at http://www.dailymail.co.uk/news/article-1145127/The-Prince-hypocrites-Charles-embarks-16-000-mile-green-crusade--aboard-private-jet.html, viewed 7/25/2009.
[413]Crichton, *State of Fear*, p. 487.
[414]Kohák, *Green Halo*, p. 8.
[415]Gore, *Earth in the Balance* (2000), p. 24.
[416]Manes, *Green Rage*, p. 240.
[417]Denis Hayes and Daryl Hannah, "Fame Exchange," *Delta Sky Magazine*, April 2009, p. 69.
[418]Mother Earth, *Delta Sky Magazine*, (ed.) Jane Haugen Olson, April 2010.
[419]Jacquet, Karine, SOS ici la Terre, *Air France Magazine*, June 2009, p. 165.
[420]Photos of Earth Suffering, in (ed.) Patrick King, Cam Graham, Um Sun-Hye, {Korean editors: Lim Ji-Hee, Lee Suk-Hee, Yoo Joo-Hee}, *Beyond*, June 2009, Vol. 33, Seoul, Korea.
[421]Going Green, *US Airways Magazine*, (ed.) Lance Elko, April 2010.
[422]Spencer, Roy, Testimony before the Senate Environment and Public Works Committee, 22 July 2008, online at http://epw.senate.gov/public/index.cfm?FuseAction=Files.View&FileStore_id=e12b56cb-4c7b-4c21-bd4a-7afbc4ee72f3, viewed 7/2/2009.
[423]Woody, Todd, Unions look for leverage on green jobs, *International Herald Tribune*, 20-21 June 2009, p. 14.
[424]Charkes, Juli S., Volunteers Work to Control Invasive Vines, *The New York Times*, 10 May 2009, p. WE2.
[425]Zoo urged to kill abandoned polar bear cub, *Herald and Weekly Times*, 21 March 2007, online at http://www.news.com.au/heraldsun/story/0,21985,21422322-663,00.html, viewed 7/7/2009.
[426]Klaus, *Blue* Planet, p. 4.
[427]Rifkin, Jeremy, *Biosphere Politics: a new consciousness for a new century*, Crown Publishers Inc., 1991, p. 285.
[428]Ehrlich, Paul R., and Anne H. Ehrlich, *The dominant animal: human evolution and the environment*, Island Press, 2008, p. 3.
[429]Diamond, Stanley, *In Search of the Primitive: A Critique of Civilization*, New Brunswick, N-J.: Transaction, 1974, p. 172.
[430]Klaus, *Blue Planet*, p. 5.
[431]Gibbon, Edward, *An Abridgement of Mr. Gibbon's History of the Decline and Fall of the Roman Empire* (second ed.), Vol. 1, Cadell and Davies, London, 1807, p.25.

[432]Klaus, *Blue Planet*, p. 4.
[433]Edwards, Jonathan, *Notes on Scripture*, in (ed.) Stephen J. Stein, *Works of Jonathan Edwards*, Volume15, Yale University Press, 1998, p. 521.
[434]Calvin, *Genesis*, p. 173.
[435]Calvin, *Genesis*, pp. 64, 65.
[436]Goklany, Indur M., *The Improving State of the World: Why We're Living Longer, Healthier, More Comfortable Lives on a Cleaner Planet*, Cato Institute, 2007.
[437]Ehrlich, *Population Bomb*, p. 150.
[438]Ehrlich and Ehrlich, *One with Nineveh*, p. 15.
[439]Spengler (David P. Goldman), Why Nations Die, *Asia Times Online*, 16 August 2005.
[440]Rifkin, Jeremy (with Ted Howard), *Entropy: A New World View* (New York: Bantam, [1980] 1981),p. 210.
[441]Wilson, *Biophilia*, p. 121.
[442]Gore, *Earth in the Balance* (2000), pp. 244-245.
[443]Singer, *Practical Ethics*, p. 278.
[444]Acts 10:13.
[445]Singer, *Practical Ethics*, p. 175.
[446]Peter Singer, *Animal Liberation: A New Ethics for our Treatment of Animals* (New York: Random House, 1975).
[447]Peter Singer, "Heavy Petting," *Nerve*, March 1 2001, online at http://www.nerve.com/opinions/singer/heavypetting, viewed 7/7/2009.
[448]Singer, *Practical Ethics*, p. 89.
[449]R. Ellis, *Monsters of the Sea* (Guilford, CT: Lyons Press, 1994), p. 245.
[450]Martin Durkin, director, *The Great Global Warming Swindle* (DVD), Wag TV, 2007.
[451]Patrick Moore, "Hard Choices for the Environmental Movement," *Oregon Wheat Magazine*, March 2000.
[452]Oliver "Buzz" Thomas, "Might our Religion Be Killing Us?" *USA Today*, April 21, 2008, republished online at http://www.applythebrakes.com/news_detail.asp?id=2, viewed 7/15/2010.
[453]Thomas, "Might Our Religion Be Killing Us?"
[454]J. Matthew Sleeth, J. *Serve God, Save the Planet: A Christian Call to Action* (Grand Rapids: Zondervan, 2001), p. 31.
[455]Goklany, "Now for the Good News."
[456]Goklany, "Now for the Good News."
[457]Li Xing, "Population Control Called Key to Deal," *China Daily*, December 10, 2009, p. 10, online at http://www.chinadaily.com.cn/china/2009-12/10/content_9151129.htm, viewed 7/15/2010.
[458]Jonathan Leake and Brendan Montague, "UK Population Must Fall to 30m, Says Porritt," *Sunday Times*, 22 March 2009, online at http://www.timesonline.co.uk/tol/news/politics/article5950442.ece, viewed 7/15/2010.
[459]Gore, *Earth in the Balance* (2000), p. 318.
[460]*Countdown to Doomsday*, Sci Fi Channel, NBC Media Productions, June 14, 2006.
[461]Paul Ehrlich, Anne Ehrlich, and John Holdren. *Ecoscience: Population, Resources, and Environment* (New York: W. H. Freeman and Company, 1977), p. 954.
[462]John P. Holdren, Gretchen C. Daily, and Paul R. Ehrlich, "The Meaning of Sustainability: Biogeophysical Aspects," in *Defining and Measuring Sustainability:*

*The Biogeophysical Foundations*, edited by Mohan Munasinghe and Walter Shearer (Washington: World Bank, 1995).
[463]Ehrlich, Ehrlich, and Holdren. *Ecoscience* , pp. 837-838.
[464]Ehrlich and Ehrlich, *One with Nineveh*, p. 79.
[465]John P. Holdren, "Science and Technology for Sustainable Well-Being," *Science*, vol. 319, no. 5862 (January 25, 2008), DOI: 10.1126/science.1153386, pp. 424-434, republished online at http://www.compete.org/images/uploads/File/ESIS%20Progressive%20Downloads/Holdren%20-%20Science%20and%20Technology%20for%20Sustainable%20Well-Being,%20Jan08%202.pdf, viewed 7/15/2010.
[466]Ehrlich, *Population Bomb*, p. 131.
[467]Manes, *Green Rage*, p. 34.
[468]Ehrlich and Ehrlich, *One with Nineveh*, p. 185.
[469]Ehrlich, *Population Bomb*, p. 44.
[470]John P. Holdren, *Population and the American Predicament: The Case Against Complacency* (Cambridge, MA: Daedalus, MIT Press, 1973), pp. 31-34.
[471]J. P. Holdren and R. K. Pachauri, "Energy," in *An Agenda of Science for Environment and Development into the 21st Century*, edited by J. C. I. Dooge, et al. (Cambridge: Cambridge University Press, 1992), p. 103.
[472]Spengler (David P. Goldman), "They Made a Democracy and Called It Peace," *Asia Times Online*, March 8, 2005, online at http://www.atimes.com/atimes/Front_Page/GC08Aa02.html, viewed 7/15/2010.
[473]Simon, *Ultimate Resource 2*, p. 548.
[474]Ashley Thorne and Peter Wood, "Sustainabullies," *National Association of Scholars*, February 12, 2009, online at http://www.nas.org/polArticles.cfm?Doc_Id=551, viewed 2/14/2009.
[475]World Commission on Environment and Development, *Our Common Future: The Brundtland Report*, Oxford University Press, 1987, p. 43.
[476]Holdren, Daily, and Ehrlich, "Meaning of Sustainability: Biogeophysical Aspects," in *Defining and Measuring Sustainability: The Biogeophysical Foundations*, ed. Munasinghe and Shearer.
[477]Gerd Leipold, interview on BBC's *Hardtalk*, online at http://news.bbc.co.uk/2/hi/programmes/hardtalk/8184392.stm, viewed 26/8/2009.
[478]Fowler, *Greening of Protestant Thought*, p. 147.
[479]Fowler, *Greening of Protestant Thought*, p. 73.
[480] Strong, *Where on Earth are We Going?* p. 237.
[481]Klaus, *Blue* Planet, p. 1.
[482]Strong, *Where on Earth are We Going?* p. 225.
[483]Strong, *Where on Earth are We Going?* p. 237.
[484]William J. Clinton, Executive Order 12852, online at http://www.archives.gov/federal-register/executive-orders/pdf/12852.pdf, viewed 7/15/2010.
[485]Gore, *Earth in the Balance* (2000), p. 225.
[486]John Calvin, *Commentary on the Book of Psalms*, translated by James Anderson, in *Calvin's Commentaries*, 22 vols. (Grand Rapids: Baker, 1984 reprint), vol. 4, part 2, p. 190.
[487]Denis Hayes and Daryl Hannah, "Fame Exchange," *Delta Sky Magazine*, April 2009, p. 68.

[488]Neddy Astudillo, et al., "God's Earth is Sacred: An Open Letter to Church and Society in the United States," February 4, 2005, online at http://www.ncccusa.org/news/14.02.05theologicalstatement.html, viewed 7/29/2009.
[489]Thomas Sieger Derr, "Environmental Theologies," *First Things*, Friday, 24 April 2009, online at http://www.firstthings.com/onthesquare/2009/04/environmental-theologies, viewed 7/15/2010.
[490]Longman, *The Empty Cradle*, p. 34.
[491]Xan Rice, "High Birthrate Threatens to Trap Africa in Cycle of Poverty," *The Guardian*, 1 September 2006, online at http://www.guardian.co.uk/theguardian/2006/sep/01/guardianweekly.guardianweekly1, viewed 7/15/2010.
[492]Durkin, *Great Global Warming Swindle* (DVD).
[493]Machen, *Christianity and Liberalism*, p. 123.
[494]"Remarks by the President on Jobs, Energy Independence, and Climate Change," East Room of the White House January 26, 2009, online at http://www.whitehouse.gov/blog_post/Fromperiltoprogress/, viewed 8/3/2009.
[495]Direct Testimony of B. Keith Trent, (p. 7), North Carolina Utilities Commission, Docket No. E-7, Sub 909, Volume 1, 2 June 2009, p. 8.
[496]Direct Testimony of Dhiaa M. Jamil, (p. 14), North Carolina Utilities Commission, Docket No. E-7, Sub 909, Volume 1, 2 June 2009, p. 99.
[497]Direct Testimony of B. Keith Trent, (p. 12), North Carolina Utilities Commission, Docket No. E-7, Sub 909, Volume 1, 2 June 2009, p. 13.
[498]Direct Testimony of B. Keith Trent, (p.7), North Carolina Utilities Commission, Docket No. E-7, Sub 909, Volume 1, 2 June 2009, p. 8.
[499]David Zahniser and Phil Willon, "DWP Rates May Rise Between 8% and 28% to Pay for Mayor's Green Initiatives," *Los Angeles Times*, March 15, 2010, online at http://www.latimes.com/news/local/la-me-dwp-rates16-2010mar16,0,7870063.story, viewed 3/15/2010.
[500]Mariam Isa, "South Africa: Eskom's Call to Country—Use 20 Percent Less Power," *Business Day*, January 21, 2008, online at http://allafrica.com/stories/200801210809.html, viewed 7/31/2009.
[501]"Eskom: Tariff Hike for the Good of the Country," *Mail and Guardian*, June 8, 2009, online at http://www.mg.co.za/article/2009-06-08-eskom-tariff-hike-for-the-good-of-country, viewed 7/31/2009.
[502]The 2010 FIFA World Cup will be held in South Africa, and there has been serious concern that the organizers cannot guarantee a continuous power supply during games.
[503]Bryan Walsh, "Cash for Clunkers: How Big an Environmental Boost?" *Time Magazine*, August 5, 2009, online at http://www.time.com/time/health/article/0,8599,1914602,00.html, viewed 7/15/2010.
[504]Sindya N. Bhanoo, "Fuel Taxes Must Rise, Harvard Researchers Say," Dot Earth, *New York Times*, March 2, 2010, online at http://dotearth.blogs.nytimes.com/2010/03/02/fuel-taxes-must-rise-harvard-researchers-say/, viewed 3/2/2010.
[505]Amanda Little, "Tim Wirth Says Imperfect Deal at Copenhagen Better than No Deal," December 17, 2009, online at http://www.grist.org/article/2009-12-17-tim-wirth-says-imperfect-deal-at-copenhagen-better-than-no-deal/, viewed 12/17/2009.

[506]Christopher Martin and Mario Parker, "Wind Promises Blackouts as Obama Strains Grid With Renewables," *Bloomberg*, August 7, 2009, online at http://www.bloomberg.com/apps/news?pid=washingtonstory&sid=arbHcz0ryM_E, viewed 8/7/2009.
[507]Paul R. Ehrlich and Anne H. Ehrlich, *The End of Affluence* (New York: Ballantine Books, 1980), p. 190.
[508]Cass R. Sunstein, *After the Rights Revolution: Reconceiving the Regulatory State* (Cambridge, MA: Harvard University Press, 1993), p. 69.
[509]Cass R. Sunstein and Martha C. Nussbaum, *Animal Rights: Current Debates and New Directions* (Oxford: Oxford University Press, 2004), p. 11.
[510]Christopher Neefus, "In the 70s, Obama's Science Adviser Endorsed Giving Trees Legal Standing to Sue in Court," *Cybercast News Service*, 30 July 2009, online at http://www.cnsnews.com/public/content/article.aspx?RsrcID=51756, viewed 7/30/2009.
[511]Laurence H. Tribe, "Ten Lessons Our Constitutional Experience Teaches Us about the Puzzle of Animal Rights: The Work of Steven M. Wise," *Animal Law Review*, vol. 7, no. 1 (2001).
[512]Helena Bachmann, "Lawyers for Animals? Up for a Vote in Switzerland," *Time Magazine*, February 12, 2010, online at http://www.time.com/time/world/article/0,8599,1963933,00.html, viewed 3/2/2010.
[513]Joanne Nova, *Climate Money* (Washington: Science and Public Policy Institute, July 21, 2009), p. 15, online at http://scienceandpublicpolicy.org/images/stories/papers/originals/climate_money.pdf, viewed 7/15/2010.
[514]E. Jay Donovan, "Energy Price Schemes Helped Doom Bailout Firms," *Environment and Climate News*, January 2009, The Heartland Institute, p. 13, online at http://www.heartland.org/full/24371/Energy_Price_Schemes_Helped_Doom_Bailout_Firms.html, viewed 7/15/2010.
[515]James Hansen, "G-8 Failure Reflects U.S. Failure on Climate Change," *Huffington Post*, 9 July 2009, online at http://www.huffingtonpost.com/dr-james-hansen/g-8-failure-reflects-us-f_b_228597.html, viewed 7/13/2009.
[516]"Henry Waxman has a Plan ... for Your Living Room, Showerhead, Jacuzzi ...," *The Wall Street Journal*, Review and Outlook, April 10, 2009, p. A12, online at http://online.wsj.com/article/SB123933057062907775.html, viewed 7/15/2010.
[517]Gaia Vince, "One Last Chance to Save Mankind," *New Scientist*, Issue 2692, 23 January 2009, online at http://www.newscientist.com/article/mg20126921.500-one-last-chance-to-save-mankind.html?full=true (subscription required), viewed 7/15/2010.
[518]Bret Stephens, "Global Warming and the Poor," *The Wall Street Journal*, 4 August 2009, p. A11, online at http://online.wsj.com/article/SB10001424052970204313604574327992553917308.html, viewed 7/15/2010.
[519]Cited by Simon, *Ultimate Resource 2*, p. 573.
[520]Susan Easton, "The Church of England Invests in Al Gore," *Human Events*, 14 January 2009, online at http://www.humanevents.com/article.php?id=30269, viewed 2/12/2009.
[521]Nova, *Climate Money*. p. 2.

[522]John Tierney, "E-Mail Fracas Shows Peril of Trying to Spin Science," *New York Times*,1 December 2009, p. D1, online at http://www.nytimes.com/2009/12/01/science/01tier.html, viewed 7/15/2010.
[523]Richard S. Lindzen, "Resisting Climate Hysteria," *Quadrant Online*, 26 July 2009, online at http://www.quadrant.org.au/blogs/doomed-planet/2009/07/resisting-climate-hysteria, viewed 8/8/2009.
[524]S. Fred Singer, *Hot Talk, Cold Science, Global Warming's Unfinished Debate*, 2d ed. S. (Oakland, CA, and Washington, DC: Independent Institute, 1999), frontispiece.
[525]Extracts of Gore's Nobel Prize for Peace Lecture on December 10, 2007, are online at http://nobelprize.org/nobel_prizes/peace, viewed 5/3/2008.
[526]Al Gore, *Earth in the Balance: Forging a New Common Purpose*, rev. ed. (London: Earthscan Ltd., 2007), pp. 220, 221.
[527]Ruether, *Gaia and God*, p. 58.
[528]Passmore, *Man's Responsibility for Nature*.
[529]Gore, *Earth in the Balance* (2000), p. 258.
[530]Charles T. Rubin, *The Green Crusade* (New York: Rowman & Littlefield, 1994), p. 176.
[531]Thomas Berry, "Into the Future," in *This Sacred Earth*, ed. Gottlieb, p. 410.
[532]Mark R. Stoll, *Protestantism, Capitalism, and Nature in America* (Albuquerque: University of New Mexico Press, 1997), pp. 13-28; Fowler, *Greening of Protestant Thought*, pp. 62-64.
[533]Singer, *Practical Ethics*, p. 22.
[534]Gary North, *Unholy Spirits: Occultism and New Age Humanism* (Tyler, TX: Dominion Press, 1986), p. 30.
[535]See 2 Timothy 3:16 and 1 Timothy 4:1.
[536]Jonathan Edwards, *The Works of President Edwards*, vol. 8 (London: James Black & Son, 1817), p. 226.
[537]Dewey J. Hoitenga, *Faith and Reason from Plato to Plantinga: An Introduction to Reformed Epistemology* (Albany: State University of New York Press, 1991).
[538]Stuart H. Merriam, *Paul the Apostle: At the Edge by Faith* (n.p.: Fenestra Books, 2004), pp. 106-108.
[539]Gore, *Earth in the Balance* (2000), p. 264.
[540]Gore, *Earth in the Balance* (2000), p. 256.
[541]Lynn Townsend White, Jr, "The Historical Roots of Our Ecologic Crisis," *Science*, vol. 155, no. 3767 (March 10, 1967), pp 1203–1207.
[542]David Kinsley, "Christianity as Ecologically Harmful" and "Christianity as Ecologically Responsible," in *This Sacred Earth*, ed. Gottlieb, pp. 104, 124.
[543]Janetta Rebold Benton, *Holy Terrors: Gargoyles on Medieval Buildings* (New York: Abbeville Press, 1997), p. 23.
[544]Gary R. Varner, *Gargoyles, Grotesques & Green Men: Ancient Symbolism in European & American Architecture* (Lulu.com, 2008), p. 44.
[545]Wendell Berry, *Christianity and the Survival of Creation* (New York: Association for Religion and Intellectual Life, 1993).
[546]J. Gresham Machen, "Jesus and Paul," in *Biblical and Theological Studies by the Members of the Faculty of Princeton Theological Seminary* (New York: Charles Scribner's Sons, 1912), p. 565.
[547]Jonathan Edwards, *Original Sin*, in *Works*, vol. 3, p. 279.

[548]Stoll, *Protestantism, Capitalism, and Nature in America*, pp. 13-24.
[549]Peter Singer, "Eating Ethically," *Free Inquiry*, vol. 25, no. 4 (June-July, 2005), pp. 18-19. Philosophical vegetarianism was stronger in ancient Greece and Rome: it was supported by Pythagoras, Empedocles, Plutarch, Plotinus, Porphyry, and, in some passages, Plato.
[550]T. H. Jukes, "Organic Apple Juice No Antidote for Alar," *J. Am. Dietetic Assoc.*, 90(3) (1990), p. 371.
[551]Hikoya Hayatsu, *Mutagens in Food Detection and Prevention* (New York: CRC Press, 1991).
[552]Virginia Postrel, "Of Mice and Men: Bruce Ames Interview," *Reason*, November 1994, online at http://www.reason.com/news/show/32261.html, viewed 8/14/2009.
[553]Dick Taverne, "The Costly Fraud that is Organic Food," *The Guardian*, 6 May 2004, online at http://www.guardian.co.uk/lifeandstyle/2004/may/06/foodanddrink.organics/print, viewed 6/6/2005.
[554]Ruether, *Gaia and God*, p. 261.
[555]Ruether, *Gaia and God*, p. 225.
[556]David Gilson, "Chew the Right Thing: The Philosopher Talks About Ethical Eating, Fast-Food Burritos, and Why Local Food Is Overhyped," *Mother Jones*, May 3, 2006, online at http://www.motherjones.com/politics/2006/05/chew-right-thing, viewed 10/8/2009.
[557]Peter Singer, "Animal Liberation and Vegetarianism," in *The Ethics of Food: A Reader for the Twenty-First Century*, edited by Gregory E. Pence (New York: Rowman & Littlefield, 2002), p. 43.
[558]Peter Singer, "Peter Singer FAQ," Princeton University, online at http://www.princeton.edu/~psinger/faq.html, viewed 10/8/2009.
[559]Kohák, *Green Halo*, p. 41.
[560]Kohák, *Green Halo*, p. 42.
[561]Laurie Goodstein, "At Home with Jim Ball; Living Day to Day by a Gospel of Green," *The New York Times*, March 8, 2007, online at http://query.nytimes.com/gst/fullpage.html?res=9C06E0D71531F93BA35750C0A9619C8B63&sec=&spon=&pagewanted=1, viewed 7/17/2010.
[562]James Hansen, "Interview with Dr. James Hansen, Director of NASA's Goddard Institute for Space Studies," *Supreme Master TV*, 12 January 2009, online at http://www.suprememastertv.com/bbs/board.php?bo_table=featured&wr_id=215&url=link2_0#v, viewed 11/8/2009.
[563]Julie Jowit, "UN Says Eat Less Meat to Curb Global Warming," *The Observer*, September 7, 2008, online at http://www.guardian.co.uk/environment/2008/sep/07/food.foodanddrink, viewed 7/18/2010.
[564]Holdren and Pachauri, "Energy," in *An Agenda of Science*, ed. Dooge et al., pp. 103-118.
[565]Peter Singer, "Vegetarianism," in *The Oxford Companion to Philosophy*, edited by Ted Honderich (Oxford, UK: Oxford University Press, 1995), p. 897.
[566]Sarah McFarland Taylor, "Reinhabiting Religion: Green Sisters , Ecological Renewal, and the Biogeography of Religious Landscape," in *This Sacred Earth*, ed. Gottlieb, p. 622.

[567]Sarah McFarland Taylor, *Green Sisters: A Spiritual Ecology* (Cambridge, MA: Harvard University Press, 2007), p. 184.
[568]Marina Frolova-Walker, "National in Form, Socialist in Content: Musical Nation-Building in the Soviet Republics," *Journal of the American Musicological Society*, 51 (1998), pp. 331-371.
[569]Robert N. Proctor, *The Nazi War on Cancer* (Princeton: Princeton University Press, 2000).
[570]Petteri Pietikainen, "The Volk and Its Unconscious: Jung, Hauer and the 'German Revolution,'" *Journal of Contemporary History*, 35:4 (2000), DOI: 10.1177/002200940003500401, pp. 523-539.
[571]Jonah Goldberg, *Liberal Fascism: The Secret History of the American Left, From Mussolini to the Politics of Meaning* (New York: Doubleday, 2008), p. 19.
[572]Robert Proctor, "The Anti-Tobacco Campaign of the Nazis: A Little-Known Aspect of Public Health in Germany, 1933-45," *British Medical Journal*, vol. 313, no. 7070 (1996), pp. 1450-1453.
[573]Biehl and Staudenmaier, *Ecofascism*, p. 33.
[574]Peter Singer, "All Animals Are Equal," in *Animal Rights and Human Obligations*, edited by Tom Regan and Peter Singer (Upper Saddle River, NJ: Prentice Hall, 1989), pp. 148-162.
[575]D. L. Amaru and C. J. Field, "Conjugated Linoleic Acid Decreases MCF-7 Human Breast Cancer Cell Growth and Insulin-like Growth Factor-1 Receptor Cells, Lipids," *PMID* 44:5 (May 2009), pp. 449-458.
[576]Neil Mitchell, "Food Nazis May Choke But Burgers Get the Green Light," *Herald Sun*, July 30, 2009.
[577]Thomas Sowell, "Greed, Gravity and Obama Sermonettes," *WorldNetDaily*, August 11, 2009, online at http://www.wnd.com/index.php?fa=PAGE.view&pageId=106560, viewed 11/8/2009.
[578]Fowler, *Greening of Protestant Thought*, p. 147.
[579]"Time to Banish the God of Growth," *New Scientist*, 2678, October 15, 2008, online at http://www.newscientist.com/article/mg20026782.500-editorial-time-to-banish-the-god-of-growth.html, viewed 7/3/2009.
[580]Al Gore, "Comments to the Smith School of Enterprise and the Environment," Oxford University, online at http://oxforddigital.tv/streaming/algore0709.php, viewed 10/8/2009.
[581]Gore, *Earth in the Balance* (2000), p. 303.
[582]Minute on global warming and climate change, GEN/PUB 05, Central Committee of the World Council of Churches, Geneva, Switzerland, 13-20 February 2008, , p. 5610, online at http://www.oikoumene.org/en/resources/documents/central-committee/geneva-2008/reports-and-documents/public-issues/minute-on-global-warming-and-climate-change.html, viewed 7/7/2009.
[583]Jeffrey Folks, "My Socialist Past," *American Thinker*, March 15, 2009, online at http://www.americanthinker.com/printpage/?url=http://www.americanthinker.com/2009/03/my_socialist_past.html, viewed 8/13/2009.
[584]Friedrich August Hayek, *The Constitution of Liberty* (Chicago: University Of Chicago Press, 1960, p. 87.

[585]Bruce Horovitz, "Makers of Plastic Bags to Use 40% Recycled Content by 2015," *USA Today*, April 20, 2009, online at http://www.usatoday.com/money/industries/environment/2009-04-20-plastic-bags-recyclable_N.htm, viewed 8/14/2009.
[586]Joel McDermon, *God Versus Socialism: A Biblical Critique of the New Social Gospel* (Powder Springs, GA: American Vision Press, 2009.
[587]Gary North, "The Bible Mandates Free Market Capitalism. It Is Anti-Socialist. The Proof Is Here: 9,000 Pages of Exposition, Verse by Verse," online at http://www.garynorth.com/public/department57.cfm, viewed 2/6/2009.
[588]Robinson, *Saving God's Green Earth*.
[589]Tri Robinson, "Please Forgive Us," *Huffington Post*, December 15, 2008, online at http://www.huffingtonpost.com/tri-robinson/please-forgive-us_b_151231.html, viewed 8/13/2009.
[590]Willis J. Jenkins, *Ecologies of Grace: Environmental Ethics and Christian Theology* (Oxford: Oxford University Press, 2008).
[591]Seth Borenstein, "Many Call '350' Key for Climate Pact," *San Francisco Chronicle*, December 14, 2009, p. A2.
[592]Gore, *Earth in the Balance* (2000), p. 248.
[593]Extracts of Gore's Nobel Prize for Peace Lecture on December 10, 2007, are online at http://nobelprize.org/nobel_prizes/peace, viewed 5/3/2008.
[594]Klaus, *Blue Planet*, p.6.
[595]Bill McKibben, *Deep Economy: The Wealth of Communities and the Durable Future* (New York: Times Books, 2007), p. 22.
[596]Sleeth, *Serve God, Save the Planet*, p. 38.
[597]Jeff Jacoby, "A World Full of Good News," *The Boston Globe*, May 13, 2007, online at http://www.boston.com/news/globe/editorial_opinion/oped/articles/2007/05/13/a_world_full_of_good_news/, viewed 7/18/2010.
[598]U.S. Agency for International Development, "What Is Food for Peace?" online at http://www.usaid.gov/our_work/humanitarian_assistance/ffp/, viewed 4/5/2010.
[599]John De Graaf, David Waan, and Thomas H. Naylor, *Affluenza: The All-Consuming Epidemic* (San Francisco: Barrett-Koehler, 2001), p. 13.
[600]Edward P. Echlin, *Live Simply: Let Others Live* (London: Progressio, 2007), p. 5.
[601]D. Martyn Lloyd-Jones, *Studies in the Sermon on the Mount* (Grand Rapids: Eerdmans Publishing, 1976), p. 29.
[602]Sleeth, *Serve God, Save the Planet*, p. 18.
[603] McKibben, *Deep Economy*, p. 167.
[604]Patrick McIlheran, "Eating Too Much," *New York Sun*, May 23, 2008.
[605]Singer, *Practical Ethics*, pp. 283, 285, and 287.
[606]Monckton, "Great Is Truth."
[607]Evo Morales, "Let's Respect Our Mother Earth," Letter from President Evo Morales to the member representatives of the United Nations on the issue of the environment, September 27, 2007, online at http://www.zcommunications.org/znet/viewArticle/14365, viewed 8/8/2009.
[608]Gore, "Comments to the Smith School."
[609]Speech by Mr. Jacques Chirac, French President, to the VIth Conference of the Parties to the United Nations Framework Convention on Climate Change, The Hague, November 20, 2000, online at www.elysee.fr, viewed 7/10/2009.

[610]"Harper Letter Called Kyoto 'Socialist Scheme,'" *The Toronto Star*, 30 January 2007, online at http://www.thestar.com/article/176382, viewed 7/17/2010.
[611]Steven Erlanger, "Red to Green: Riding E.U. Ecology Wave," *International Herald Tribune*, June 22, 2009, p. 3, online at http://www.highbeam.com/doc/1P1-165101640.html (subscription required), viewed 7/17/2010.
[612]Tom Athanasiou, *Divided Planet: The Ecology of Rich and Poor* (Athens: University of Georgia Press, 1998), p. 3.
[613]Eliza Strickland, "The New Face of Environmentalism," *East Bay Express*, November 2, 2005, online at http://www.eastbayexpress.com/ebx/PrintFriendly?oid=290098, viewed 8/11/2009.
[614]Eric D. Weitz, *A Century of Genocide: Utopias of Race and Nation* (Princeton: Princeton University Press, 2003), p. 113.
[615]*Jonestown: The Life and Death of Peoples Temple*, DVD, (dir.) Stanley Nelson, PBS Paramount, 2006, transcript online at http://www.pbs.org/wgbh/americanexperience/jonestown/transcript/4/, viewed 8/19/2009.
[616]Fowler, *Greening of Protestant Thought*, p. 146.
[617]McKibben, *Deep Economy*, p. 94.
[618]Philip Pullella, "Rich Nations Must Assume Environmental Duties: Pope," *Reuters*, December 15, 2009, online at http://uk.reuters.com/article/idUKTRE5BE1UR20091215, viewed 7/17/2010.
[619]Thomas Watson, *The Art of Divine Contentment*, in *A body of Practical Divinity consisting of above 176 sermons on the Lesser Catechism composed by the Reverend Assembly of Divines at Westminster with a supplement of some sermons on several texts of scripture together with The Art of Divine Contentment to which is added Christ's various Fulness* (Glasgow: Glasgow College, Archibald Ingram, James Dichman, John Hamilton, John Glassford, 1741), p. 656.
[620]Watson, *Art of Divine Contentment*, p. 13.
[621]"Move over Tiger: N. Korea's Kim Shot 38 Under Par His 1st Time Out," special to *World Tribune.com*, June 16, 2004, online at http://www.worldtribune.com/worldtribune/WTARC/2004/ea_nkorea_06_16.html, viewed 8/15/2009.
[622]"The Odds on Making a Hole-in-One," *Golf Digest via Bnet*, March 2000, online at http://findarticles.com/p/articles/mi_m0HFI/is_3_51/ai_59554906/, viewed 6/16/2009.
[623]Angie Knox, "Profile: Kim Jong-il," June 9, 2000, online at http://news.bbc.co.uk/2/hi/asia-pacific/783967.stm, viewed 10/6/2009.
[624]Chol-Hwan Kang and Pierre Rigoulot, *The Aquariums of Pyongyang* (New York: Basic Books, 2001), p. 82.
[625]"North Korea Denounces Global Warming," *WorldNet Daily*, March 26, 2004, online at http://www.wnd.com/news/article.asp?ARTICLE_ID=37756, viewed 10/6/2009.
[626]Biehl and Staudenmaier, *Ecofascism*, p. 26.
[627]Gore, *Earth in the Balance* (1992), p. 239.
[628]Klaus, *Blue Planet*, p.78.
[629]Phillips Payson, "Massachusetts Election Sermon of 1778," in *Pulpit of the American Revolution*, ed. Thornton, p. 339.
[630]The Westminster Presbyterian, *Religion and Society*, online at http://www.westminsterconfession.org/identity/religion-and-society.php, viewed 8/20/2009.

[631]Weisskopf, "Energized by Pulpit or Passion," p. A1.
[632]James Stuart, *A History of the Zulu Rebellion 1906: and of Dinuzulu's Arrest, Trial, and Expatriation* (New York: Macmillan, 1913), p. 107.
[633]John A. Chalmers, *Tiyo Soga: A Page of South African Mission Work* (London: Hodder and Stoughton, 1878), p. 115.
[634]Roger B. Beck, *The History of South Africa*, The Greenwood Histories of the Modern Nations (Santa Barbara, CA: Greenwood, 2000), p. 57.
[635]Machen, *Christianity and Liberalism*, p. 178.

# Index


## A

Abortion 38, 42, 65, 154, 167, 220, 230
Adam 39, 81, 93, 94, 97, 98, 99, 100, 103, 104, 106, 107, 112, 133, 135, 141, 143, 144, 145, 146, 147, 148, 156, 167, 170, 176, 177, 178, 179, 208
Affluenza 257
African Methodist Episcopal Church 64
Agenda 21 225
Air France 202
A Nation at Risk 49
Angels 89, 106, 110, 111, 112, 139, 146, 178, 184
Animal 146
  value of 216
Anthropocentrism 73, 78, 119, 128, 241, 246, 249
Ark of the Covenant 62, 129, 130, 131, 132
Arndt, Ernst 72
Ashtoreth 130
Astarte 130
Atheism 123
  end of 137
Au Sable Institute 67, 70

## B

Bahro, Rudolf 77
Baige, Zhao 219
Ball, Jim
  Evangelical Environmental Network 250
Barna 57
Barnes, Albert 94
Beisner, E. Calvin 86
Berry, Thomas 71, 72, 241
Berry, Wendell 160, 163, 165, 167, 244
Bible
  1 Chronicles 16:33 128
  1 Corinthians 1:25 101
  1 Corinthians 2:2 51
  1 Corinthians 3:16–17 132
  1 Corinthians 3:17 247
  1 Corinthians 4:10 269
  1 Corinthians 6:19 135, 247
  1 Corinthians 6:19–20 173
  1 Corinthians 7:4 177
  1 Corinthians 8:13 249
  1 Corinthians 11 111
  1 Corinthians 11:7 106
  1 Corinthians 11:9 178
  1 Corinthians 15:22 177
  1 Corinthians 15:24,25 181
  1 Corinthians 15:33 243
  1 Corinthians 15:39 93
  1 Corinthians 15:45,49 106
  1 Corinthians 15:47 99
  1 John 4:16 200
  1 John 5:3 99
  1 Kings 8:27 88, 101, 130
  1 Peter 1:14 162
  1 Peter 1:24,25 136
  1 Peter 2:5 132
  1 Peter 2:6–8 51
  1 Samuel 4:8 131
  1 Samuel 5:2–4 161
  1 Samuel 14:41 49
  1 Timothy 2:4 182
  1 Timothy 5:21 139
  1 Timothy 6:15,16 168
  1 Timothy 6:17 194
  2 Chronicles 2:9 208
  2 Chronicles 15:16 208
  2 Corinthians 5:4 246
  2 Corinthians 5:8 107
  2 Corinthians 10:5 78, 247
  2 Peter 1:13 246
  2 Peter 2:1–3 270
  2 Peter 2:12 94, 198
  2 Peter 3:6 214
  2 Peter 3:10–12 179
  2 Peter 3:16 158
  2 Samuel 6:2 129
  2 Timothy 2:12 181
  Acts 7:48 130

Acts 10:13 252
Acts 10:13–15 198
Acts 10:36 169
Acts 15:29 161
Acts 16:16–18 139
Acts 17 243
Acts 17:24 129, 169
Acts 17:24–25 123
Acts 17:26 141, 159
Colossians 2:20–23 251
Colossians 3:10 97
Deuteronomy 8:10–19 151
Deuteronomy 20:19,20 180
Deuteronomy 28:53 178
Deuteronomy 29:29 90
Deuteronomy 32:39 171, 199
Ecclesiastes 3:10–11 124
Ecclesiastes 3:11 96, 170
Ecclesiastes 3:16–22 110
Ecclesiastes 3:19 145
Ecclesiastes 5:11 222
Ecclesiastes 7:29 98, 112
Ephesians 1:4 106
Ephesians 1:21 172
Ephesians 3:9,10 111
Ephesians 4:24 99
Ephesians 5:10 57
Ephesians 5:24 177
Ephesians 5:28 177
Ephesians 5:29 246
Exodus 3:5 103
Exodus 9:16 199
Exodus 20 259
Exodus 21:28 146
Exodus 23:29 179
Exodus 24:13 176
Exodus 28:1 176
Exodus 30:36 129
Exodus 33:23 198
Exodus 34:13 208
Galatians 5:16–17 245
Genesis 1 90
Genesis 1:1 88
Genesis 1:2 140
Genesis 1:11 144
Genesis 1:14 140
Genesis 1:26 91, 144
Genesis 1:26–28 156, 173, 175, 183
Genesis 1:27 91
Genesis 1:28 216
Genesis 1:29 216
Genesis 1:31 208
Genesis 2:7 93
Genesis 2:8–15 208
Genesis 2:8,15 94
Genesis 2:15 177
Genesis 2:17 94
Genesis 2:19 94
Genesis 2:22 141
Genesis 2:23 93
Genesis 3 148
Genesis 3:7 198
Genesis 3:7,8 147
Genesis 3:10,11 146
Genesis 3:14 107
Genesis 3:15 148
Genesis 3:17 39
Genesis 3:18 178
Genesis 3:19 263
Genesis 3:20 97
Genesis 4:11 39
Genesis 6:9 113
Genesis 7:2 215
Genesis 7:3 214
Genesis 7:21–24 198
Genesis 7:24 26
Genesis 8:20 215
Genesis 8:22 209
Genesis 9:1–3 216
Genesis 9:2 197
Genesis 13:6 222
Genesis 13:10 222
Genesis 13:16 222
Genesis 18:20 98
Genesis 21:33 208
Genesis 47:13 26
Hebrews 1:3 169
Hebrews 1:14 139, 184
Hebrews 10:1 130
Hebrews 10:27 200
Hebrews 10:31 50, 227
Hebrews 12:14 104

Hosea 4:10 194
Isaiah 9:6 134
Isaiah 11:8 185
Isaiah 27:6 185
Isaiah 35:1–2 185
Isaiah 40:11 200
Isaiah 41:19–20 207
Isaiah 44:24 90
Isaiah 45:18 169
Isaiah 46:9,10 168
Isaiah 48:22 104, 263
Isaiah 49:16 200
Isaiah 49:23 226
Isaiah 57:20 263
Isaiah 57:21 227
Isaiah 58:13,14 226
James 1:17 98
James 1:22 58
James 2:19 228
James 4:7 270
Jeremiah 9:24 98
Jeremiah 17:9 183
Jeremiah 23:5 99
Jeremiah 51:29 199
Job 13:15 214
Job 22 90
Job 26:13 169
Job 38:7 140
Joel 2:1 198
John 1:48 198
John 3:16 105
John 4:24 245
John 5:39 57
John 6:53 128
John 6:63 169, 245
John 7:15–17 268
John 7:16,17 268
John 8:32 49
John 8:44 107
John 8:51 226
John 11:35 145
John 12:27 145
John 14:6 161
John 14:15 226
John 14:30 171
John 15:19 209
John 16:33 263
John 17 3 48
John 17:9–10 180
John 17:17 55, 161
John 18:36 171
John 19:26–27 249
Joshua 11:6 198
Judges 6:25 208
Judges 17:6 226
Leviticus 24:21 113
Luke 1:45 90
Luke 4:18–19 269
Luke 4:35 139
Luke 5:32 104
Luke 12:7 169
Luke 12:15 194
Luke 13:6–9 179
Luke 14:26 147
Luke 14:28 194
Luke 19:13 181
Malachi 3:6 171
Mark 1:3 49
Mark 2:27 178
Mark 8:35–36 258
Mark 9:2 198
Mark 10:42–43 184
Mark 11:23 198
Mark 12:25 139
Mark 13:2 132
Mark 14:38 245
Mark 16:15 182
Matthew 4:4 51
Matthew 5:13 50
Matthew 6:24 79
Matthew 6:32 194
Matthew 7:15–20 266
Matthew 7:16 179
Matthew 10:16 269
Matthew 12:11–12 139
Matthew 15:11 257
Matthew 16:25–26 173
Matthew 20:26 176
Matthew 20:25–28 175
Matthew 21:18–22 179
Matthew 21:19,20 198
Matthew 22:39 215

Matthew 23:3 262
Matthew 24:32 198
Matthew 25:15–32 176
Matthew 26 259
Matthew 28:19 158
Matthew 28:20 162
Micah 4:3 185
Micah 7:18 147
Philippians 1:21 61
Philippians 2:10,11 161
Philippians 4:11 262
Proverbs 8:36 161, 270
Proverbs 9:13–18 79
Proverbs 10:15 194
Proverbs 13:4 259
Proverbs 14:30 263
Proverbs 19:15 259
Proverbs 28:6 194
Psalm 2 102, 171
Psalm 2:10–12 44
Psalm 2:12 209
Psalm 8 184
Psalm 8: 4–5 110
Psalm 8:5,6 172
Psalm 8:6 176
Psalm 8:6–9 178
Psalm 14:1 228, 269
Psalm 19 135
Psalm 24:1 173
Psalm 27 49
Psalm 36:9 49
Psalm 39:4 161
Psalm 47:2 199
Psalm 51:5 159
Psalm 60:5 180
Psalm 66:5 199
Psalm 72:8 161
Psalm 89:35 102
Psalm 90:2 169
Psalm 103:11 128
Psalm 104:24 169
Psalm 115:3 168, 171
Psalm 115:16 184
Psalm 119:105 161
Psalm 128:3 178
Psalm 141:2 129
Psalm 145:3 168
Psalm 145:17 97, 176
Psalm 147:9 214
Revelation 2:27 199
Revelation 4:4 111
Revelation 4:11 169
Revelation 5:9 111
Revelation 7:11 112
Revelation 20:2 107
Revelation 21:9 180
Revelation 22:3 130
Romans 1:19 95
Romans 1:21–23 122
Romans 2:15 96
Romans 3:10 98
Romans 3:20–27 98
Romans 3:23 159
Romans 5:7 144
Romans 5:7,8 107
Romans 5:14 147
Romans 5:15,18 147
Romans 6:23 145
Romans 7:25 245
Romans 8:10 246
Romans 8:19 182
Romans 8:20 145
Romans 8:22 181
Romans 9:14 199
Romans 9:17 171, 200
Romans 9:20 171
Romans 9:21–23 170
Romans 9:26 112
Romans 11:22 200
Romans 11:33 176
Romans 13:4 176
Romans 14:14 250
Romans 16:19 269
Romans 20:25 98
Titus 1 243
Zechariah 2:8 180
Zechariah 4:7 198
Zephaniah 2:13,14 212
Birch, Charles 240
Blackfoot Indians 157
Blood, David 238
Body lice 121

Body of God 38, 75
Bradley, Ian
God is Green 229
Brainerd, David 159
Brown Campbell, Rev. Joan 65
Brown, Gordon 220
Brundtland Commission 223
Buddhism 138
Bulgarian brothels 35
Burroughs, Jeremiah 262

## C

Caesar 26, 175
Calvin, John 39, 91, 143, 148, 181, 185, 208, 228
Capitalism 29, 223, 252, 255
Good for the environment 196
Protestant hostility 225
Carbon dioxide 34, 37, 65, 190, 209, 239
Carbon trading 237
Cardinal Pell
Archbishop of Sydney 63
Cardin, Ben 237
Carolinas 69, 234
Carson, Rachel 202
Silent Spring 200
Central organizing principle 36, 37, 38, 42, 71, 72, 75, 78, 119, 134, 172, 207, 213, 224, 267, 268
Chesterton, G. K. 63, 101, 166
China 29, 34, 37, 117, 194, 227, 248, 263
Chirac, Jacques 260
Christian
balance 112
Christianreason 94
Christianuneducated 268
compromise 60
irrationalism 56, 259
pantheism 134, 138
realism 48, 56, 73, 78, 89, 109, 241, 245
Christianity Today 127
Christian Reformed Church 128
Christ saves sinners 86
Chronicles of Narnia 87
Chrysostom 111
Church of England 238
CIA 25
Cizik, Richard 66
Clarke, Adam 197
Clarke, Arthur C. 239
Clark, Gordon 56, 60, 144
Climate Gate 239
Cobb, John 240
Collectivism 37, 140, 150, 237, 253, 256, 260
Copenhagen 219, 233, 235, 238, 261
Cotton , John 48
Council for Christian Colleges and Universities 70
Creation
a cosmic theater 184
cosmology 138
designed for man 181
doctrine of 134, 141, 255
God destroys 208
intrinsic value 190, 213
natural balance 203
serves mankind 179
tool of dominion 180
uses of 198
without intrinsic value 180
Crichton, Michael 63, 202
Cuba 194
Curse, The 145, 146

## D

Dabney, Robert 25, 134, 149
Darwin, Charles 30, 72, 118, 119, 165
Dawkins, Richard 133, 135, 149
DDT 200, 202, 210
de Chardin, Teilhard 72
De Jonge, Eccy 75
Delta Airlines 202
Demeter 130
Democrat Party 24, 34
Derrida, Jacques 96
Derr, Thomas 229
Devall, Bill 73
Dewey, John 54, 55, 58, 75, 87

DeWitt, Calvin 67
Earth-Wise 229
Respecting Creation's Integrity 196
Diamond, Jared 212, 263
Collapse 222
Diamond, Stanley 205
DiCaprio, Leonardo 75
Diversity 213
Dominion 43, 44, 153, 155, 157, 160, 161, 165, 168, 175, 180, 183, 197, 216, 227, 253, 270
commanded by God 176
fruitfulness 176
ownership 169
power and right 169
Power to destroy 198
private property 175
sovereignty 171
Duke Power 234, 257
Dyson, Freeman 189

## E

Earth
Blue planet 189
No garden of Eden 208
Earth Charter 62, 125
Earth Council 226
Earth Day 25, 33, 202
Earth First! 31
Earth in the Balance 52, 64, 74, 155, 255
Earth Summit 64, 225
Echlin, Edward 257
Ecofeminism 75
EcoJustice Program 65
Ecotheology 71, 74, 76
Eden 28, 94, 139, 141, 150, 167, 176, 179, 181, 208, 216, 267, 270
Edwards, Jonathan 108, 112, 159, 172, 178, 180, 182, 207, 242, 246
Ehrenfeld, David
The Arrogance of Humanism 120
Ehrlich, Paul 25, 33, 41, 162, 191, 210, 252
Human Natures 157
One with Nineveh 210
Population Bomb 155
The End of Affluence 236
El Niño 28
Enron 237
Environmental Groups
Greenpeace 205, 218, 224
World Wildlife Fund 205
Environmentalism
Pagan roots 126
Environmental Protection Agency 190
Epimenides 243
Episcopal Church in the United States of America 64
Eugenics 120, 141
European Union 189
Evangelical 54
Evangelical Environmental Network 65, 69, 250
Evangelical Lutheran Church in America 64
Eve 93, 94, 97, 98, 104, 145, 147, 156, 177
Evolution 40, 60, 71, 72, 118, 120, 127, 136, 137, 140, 142, 165, 265
Extinction 214, 215

## F

False religion 109
Fish, Stanley 53
Flesh and spirit 109
Flood 26
Foucault, Michel 73
Foundation
New World 68
Pew Charitable Trusts 68
Rockefeller Brothers 68
Tides 68
Turner 68
V. Kann Rasmussen 68
W. Alton Jones 68
Fowler, Robert 191, 225, 241, 253, 262
Fox, Matthew 201
Francis of Assisi 162

## G

Gaia 31, 39, 52, 70, 73, 85, 131, 150, 163, 178, 201, 240, 266
Gallup 23
Gibbon, Edward 206
God
  love for man 106
  mercy 199
Goklany, Indur 166, 196
Goldberg, Jonah 252
Goldman, David 166. *See* Spengler; *See* Spengler; *See* Spengler
Goldman Sachs 238
Gore, Al 23, 26, 36, 37, 40, 52, 53, 56, 61, 64, 74, 76, 87, 100, 108, 134, 155, 191, 202, 213, 220, 226, 233, 238, 240, 243, 253, 256, 260, 265
Gottlieb, Roger 74
Great Britain 220
Greek Philosophy
  Epicurus 243
  Plato 108, 241
Green Bible 66
Green Congregations 66
Green Rage 39, 165
Green religion 59
Guevara, Che 158
Guilt manipulation 38, 66, 145, 249

## H

Haeckel, Ernst 71
Hansen, James 237, 238, 250, 260
Harper, Stephen 260
Hayek, Friedrich 254
Hayes, Denis 228
  Earth Day 202
Heidegger, Martin 73, 201, 226
Heschel, Abraham 126
Hess, Rudolf 252
Hitler 27, 105, 158, 174, 202, 251
Holdren, John 220, 250
  The Meaning of Sustainability 224
Holiness 49, 89, 90, 102, 110, 144, 161
Hollywood 44, 75, 124, 128, 150, 223
Honda, Soichino 30
Human hubris 118
Hume, David 29

## I

Image of God 18, 53, 63, 72, 78, 84, 93, 94, 97, 99, 103, 106, 108, 109, 110, 113, 119, 124, 134, 135, 159, 166, 181, 223, 240, 251
India 263
Indulgences 38
Isis 130

## J

Jain religion 84
Japan 195
Jeffords, James 70
Jenkins, Willis
  Ecologies of Grace 255
Judaism 70
Judas 135
Judgment 133
Justice 47, 49, 54, 61, 65, 69, 86, 97, 133, 159, 172, 223, 252, 255

## K

Kali 130
Kierkegaard, Søren 56, 73
Kim
  Dae-Jung 264
  Jong-Il 264
Klaus, Václav 189, 204, 206, 256, 266
Knut 204
Kohák, Erazim 88, 163, 201, 202, 249
Korean Air 202
Korea, North 264, 265, 300
Korea, South 195, 264
Kyoto 66, 260

## L

Leahy, Frederick S. 139
Lehman Brothers 237
Lenin, V.I. 34
Leopold, Aldo 72, 117, 149

Lewis, C. S. 150
Liberalism 77
Liberal theologians 100
Library of Congress 31
Lightbulb 37, 66, 175
Lindzen, Richard 239, 296
Lloyd-Jones, Martyn 258
Lomborg, Bjørn 23, 166, 239
Longman, Phillip
  The Empty Cradle 229
Lovelock, James 32, 237, 238
Luther, Martin 55

## M

Machen, J. Gresham 60, 77, 231, 271
Mainline 53, 55, 60, 63, 64, 71, 134, 225
Malthus, Robert 29, 30, 40
Manes, Christopher 39, 72, 118, 119, 165, 202, 221
Mankind
  alienation from nature 241
  and angels 111
  death unnatural 145
  dependent on Gaia 201
  dignity of 91, 158
  greater than angels 172
  greater than the cosmos 135
  greatness 112
  intrinsic value 78, 134
  naturally vicious 183
  not an animal 143
  precious to God 144
  search for meaning 124
    special creature 193
    superior to the rest of creation 178
Mao, Tse-Tung 27, 105, 158, 251
Marvin, Julia 27
Marxism 65
Marx, Karl 253, 256
Materialism
  Consequences of 130
Mather, Cotton 94, 182
Matrix, The 32, 246
Mattera, Philip 203
McFague, Sallie 75, 85, 86, 201
  Models of God 75
McKibben, Bill 64, 136, 138, 191, 255, 257, 259, 262
Meaning of life 51, 57, 61, 74, 92, 95, 103, 136, 137, 149, 191, 193, 243, 263
Menander 243
Middlebury College 126, 136
Milton, John 87
Monckton, Christopher 165, 260
Monty Python 145
Moon, Ban-Ki 195
Moore, Patrick
  Greenpeace 218
Mother Earth 34, 38, 40, 44, 266
Mother Nature 39, 150, 248
Muir, John 67, 72, 117
Murphy, J. G. 141
Museveni, Yoweri 230

## N

Næss, Arne 73, 117, 119, 122, 167, 240
Narcissism 153
Nash, Roderick 162
National Association of Evangelical 66
National Baptist Convention 64
National Council of Churches 64, 191, 229
Natural Resources Defense Council 254
NCC 64, 65, 66, 68, 69, 191. National Council of Churches; National Council of Churches; National Council of Churches
Nebuchadnezzar 84
New Age 76, 77
New Scientist 23, 31, 253
New world order 17, 59, 62, 213
New York Times 25, 70, 117
Nietzsche, Friedrich 29, 95, 100, 119, 246
Noah's Flood 199

Nobel Peace Prize 36, 56, 61, 76, 250, 256, 264
North, Gary 194, 242, 254
North Korea 124, 194, 195, 209, 253
NRPE
National Religious Partnership for the Environment 69, 71

## O

Obama, Barack 34, 220, 230, 234, 250, 259
Oelschlaeger, Max 61
Organic food 38, 57, 87, 131, 139, 140, 160, 168, 193, 247, 248, 252
Hitler 174
Orwell, George 209
Animal Farm 71
Ozone 24

## P

Paganism 79, 149, 154
Superstition 131
Pascal, Blaise 97, 108, 112, 145, 149
Passmore, John 77, 240
Peace 23, 54, 61, 104, 124, 184, 227, 251, 257, 260, 261, 264, 268, 269
PETA 42, 204
Pew Poll 23
Philistines
PhilistinesEarth Worship 133
Planned Parenthood 141
Pope, Carl 66
Popper, Karl 85
Population 29, 30, 33, 34, 40, 41, 54, 59, 122, 153, 205, 211, 212, 218, 219, 220, 230, 250, 270
Porritt, Jonathon 220
Poverty 30, 41
Presbyterian Church 64, 73, 165
President's Council on Sustainable Development 226
Prince Philip 33
Progressive National Baptist Convention 64
Puritan 48, 94, 139, 182

## R

Rape 31
Redemption 247
Reformation, Protestant 242
Relativism 58
Religious Coalition for Reproductive Choice 65
Republican Party 24, 34, 234
Rifkin, Jeremy 205, 213
Righteousness 97
Robinson, Tri
Saving God's Green Earth 255
Rockefeller 62, 125
Roman Catholic 70, 118, 119
Rosenzweig, Franz 86
Rousseau, Jean-Jacques 28, 29, 159
Rubin, Charles 241
Rudy, Sister Marilyn 251
Ruether, Rosemary 85, 126, 240, 248
Rushdoony, Rousas 175

## S

Sacredness
of nature 135
Sagan, Carl 89, 123, 125, 136, 138
San Francisco Chronicle 34, 234
Sanger, Margaret 141
Sartre, Jean-Paul 137
Satan 51, 106, 107, 139, 145, 146, 148, 149, 154, 171, 182, 197, 267, 270
Satan's hatred 107
Satchidananda 51
Scale of nature 139
Schaeffer, Francis 61, 89
Schlossberg, Herbert 47, 125
Schweitzer, Albert 120, 215, 285
Science 142
Shikwati, James 230
Sierra Club 66, 67, 72
Simon, Julian 166, 187, 223, 239

Sinful nature 104
Singer, Fred 239
Singer, Peter 42, 92, 95, 96, 113, 174, 201, 215, 241, 248, 252, 259
Sittler, Joseph 65
Sleeth, Matthew 219, 257, 258
Socialism 65, 184, 203, 211, 227, 253
contrasted with liberty 195
Sodom 98
Sola Scriptura 55
Soros, George 62
Sowell, Thomas 212, 253
Spangler, David 126, 140, 163
Spengler, Oswald 122, 150, 213
Spiritual hunger 38
Spurgeon, Charles 124
Stalin 27, 53, 105, 158, 251
Staudenmaier, Peter 265
Stiles, Ezra 175
Stoll, Mark 241
Strong, Maurice 62, 63, 70, 125, 223, 225, 256
Sunstein, Cass 236
Supreme Court 190, 193
Sustainability 223, 226, 227, 292, 293

# T

Tabernacle, The 129, 130, 131
TerraPass 38
The Hitchiker's Guide to the Galaxy 35
Thorne, Kip 138
Thornwell, James 103
Time Magazine 23, 34, 235
Tolle, Eckhart 76
Tribe, Laurence 236
Tucker, Mary Evelyn 88

# U

United Methodist Church 64
United Nations 62, 65, 70, 223, 238, 250
United States National Park Service 32
USSR 37, 195

# V

Value
Life 120
Virus 32, 39, 108, 217, 228
Visick, Vernon 127
Voluntary Human Extinction Movement 40

# W

Wade, Nicholas 121
Washington Post 54
Watson, Thomas 262
What Would Jesus Drive? 65
Whitefield, George 91
White, Lynn 64, 157, 162, 243
Wildavsky, Aaron 36
Wilson, E.0. 59, 127, 201, 213
Winfrey, Oprah 76, 126
Wirth, Tim 70, 235, 238
Wolf 55, 79, 83
Wood, Peter 223
Woodstock 51
World Bank 223
World Council of Churches 253, 280, 298
World Council of Religious Leaders 65
World Wide Fund for Nature (WWF) 33

# X

Xhosa 269

# Y

Y2K 27
Yodok gulag 195, 265

## Additional Resources

Cornwall Alliance
*For The Stewardship Of Creation*

The Cornwall Alliance is a coalition of clergy, theologians, religious leaders, scientists, academics, and policy experts committed to bringing a balanced Biblical view of stewardship to the critical issues of environment and development. The Cornwall Alliance fully supports the principles espoused in the Cornwall Declaration on Environmental Stewardship, and is seeking to promote those principles in the discussion of various public policy issues including population and poverty, food, energy, water, endangered species, habitat, and other related topics.

For more information, visit our website at www.CornwallAlliance.org
Email us at stewards@cornwallalliance.org
or mail to 9302-C Old Keene Mill Rd. Burke, VA 22015.

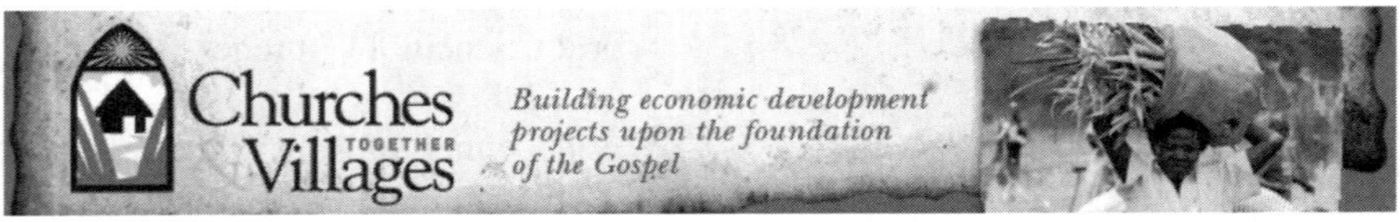


Churches and Villages Together (CVT) is a non-profit organization that works to link churches in the United States with churches and trusted Christian leaders in poor villages and other communities in developing countries. Through these partnerships, CVT coordinates evangelism, business and ethics training, and the creation of sustainable, accountable economic development projects.

Our aim is to raise both the spiritual and physical condition of needy villages, help them become self-sufficient, and see American churches receive the blessings that come from entering into relationship with their materially poor but spiritually rich brothers and sisters in Christ.

CVT began its international outreach in the East African nation of Uganda in 2008, where a pilot project is underway. CVT is now presenting its work to Christian leaders and potential church partners throughout the United States.

To begin sponsoring today, visit our website at www.ChurchesAndVillages.org.

## Additional Resources

DVD Series

**Resisting the Green Dragon** takes its cue from James 4:7, "Submit therefore to God. Resist the devil and he will flee from you." Learn how the Bible powerfully confronts environmental fears and how—in God's wise design—people and nature can thrive together.

This explosive series contains 12 half-hour sessions, along with a special 30-minute bonus documentary, and a discussion guide with questions and suggested practical applications **suitable for churches, Sunday schools, families, classrooms, home schools, and small groups.**

### TITLE and SPEAKERS

The False World View of the Green Movement — *Dr. E. Calvin Beisner*

Rescuing People from the Cult of the Green Dragon — *Dr. Peter Jones*

Logos vs. Mysticism: Environmentalism's Flight from Reason — *Dr. Vishal Mangalwadi*

From *Captain Planet* to *Avatar*: The Seduction of Our Youth — *Dr. Michael Farris*

A Brief History of Environmental Exaggerations, Myths and Downright Lies — *Dr. Steven Hayward*

Putting Out the Dragon's Fire on Global Warming — *Dr. David Legates*

How "Going Green" Impoverishes You, Your Church, and Your Society — *Hon. Becky Norton Dunlop*

Ravaging the World's Poor — *Dr. James Tonkowich*

The Green Face of the Pro-Death Agenda: Population Control, Abortion and Euthanasia — *Dr. Charmaine Yoest*

Threats to Liberty and the Move Toward a Global Government — *Dr. E. Calvin Beisner*

A Biblical Guide to Genuine Creation Stewardship — *Dr. James Tonkowich*

Go Therefore and Make Disciples: Advancing the Gospel in a World Permeated by Environmentalism — *Dr. Peter Jones*

### SPECIAL APPEARANCES

David Barton, WallBuilders
Bryan Fischer, American Family Association
Pastor Jack Hibbs, Calvary Chapel Chino Hills
Bishop Harry Jackson, Hope Christian Church
Dr. Richard Land, Southern Baptist ERLC
Tom Minnery, Focus on the Family
Dr. David Noebel, Summit Ministries
Janet Parshall, National Radio Host
Tony Perkins, Family Research Council
Dr. Frank Wright, National Religious Broadcasters
Wendy Wright, Concerned Women for America

To order, visit our website at www.ResistingTheGreenDragon.com or call us at 703-569-4653.